A RETURN TO TASMANIA

Batty Folklore; The Alcester Seal; Mystery Mongooses; Weird Weekend 2015; Another encounter with Sligo's Salamander?

The Journal of the Centre for Fortean Zoology   #54

Contents

2. Contents
3. Faculty
4. Editorial
7. Newsfile: New and Rediscovered
14. Newsfile: Chupacabras
15. Newsfile: Blue dogs
16. Newsfile: Man beasts
20. Newsfile: Mystery cats
23. Newsfile: Aquatic monsters
26. CARL MARSHALL'S COLUMN:
The Alcester Seal and other unexpected aquatic guests
33. CARL PORTMAN'S FORTEAN INVERTEBRATES
Gone Batty
37. Watcher of the Skies by Corinna Downes
48. Tasmania Expedition Report by Lars Thomas
56. A little known British pseudocryptid: The Mongoose by Richard Muirhead
68. Duties for regional representatives— a discussion document
69. Letters to the Editor
72. Book Reviews
78. Weird Weekend 2015
82. Recent books from CFZ Press

Typeset by Jonathan Downes,
Cover and Layout by SPiderKaT for CFZ Communications
Using Microsoft Word 2000, Microsoft Publisher 2000, Adobe Photoshop CS.
First published in Great Britain by CFZ Press

CFZ Press, Myrtle Cottage, Woolsery, Bideford, North Devon, EX39 5QR

© CFZ MMXV

All rights reserved. Without limiting the rights under copyright reserved above, no part of this publication may be reproduced, stored in or introduced into a retrieval system, or transmitted, in any form of by any means (electronic, mechanical, photocopying, recording or otherwise), without the prior written permission of both the copyright owners and the publishers of this book.

ISBN: 978-1-909488-39-7

Faculty of the Centre for Fortean Zoology

Hon. Life President: Colonel John Blashford-Snell

Director: Jonathan Downes
Deputy Director: Graham Inglis
Administrative Director: Corinna Downes
Zoological Director: Richard Freeman
Deputy Zoological Director: Max Blake
Technical Director: David Braund-Phillips

American Office: Nicholas Redfern, Naomi West, Richie West
Canadian Office: Robin Pyatt Bellamy
Australian Office: Rebecca Lang and Mike Williams
New Zealand Office: Tony Lucas

eBooks: Dr Andrew May
Director's Assistant: Jessica Taylor

CFZ Blog Network
Editor: Jon Downes
Sub Editor: Liz Clancy
Newsblog Editor: Corinna Downes
Regional News: Andrew May

BOARD OF CONSULTANTS
Cryptozoology Consultant: Dr Karl Shuker
Zoological Consultant: Lars Thomas
Palaeontological Consultant: Dr Darren Naish
Surrealchemist in Residence: Tony "Doc" Shiels
Ichthyological Consultant: Dr Charles Paxton
Folklorist: Jeremy Harte
Mythologist: Ronan Coghlan

REGIONAL REPRESENTATIVES
Co-ordination: Ronan Coghlan

England

Cambridgeshire: Neal McKenna
Cheshire: Richard Muirhead, Glen Vaudrey
Cumbria: Brian Goodwin
Dorset: Jonathan McGowan
Co. Durham: Dave Curtis, Jan Edwards
Greater London: Rachel Carthy, Neil Arnold
Greater Manchester: Liz Clancy
Hampshire: Darren Naish
Kent: Neil Arnold
Lancashire: Liz Clancy
Leicestershire: Mike Playfair
Merseyside: Lee Walker
Oxfordshire: Carl Portman
Staffordshire: Shoshannah Hughes
Suffolk: Matt Salusbury
Surrey: Nick Smith
West Midlands: Dr Karl Shuker,
Wiltshire: Matthew Williams
Yorkshire: Steve Jones

Wales
Gavin Lloyd Wilson, Gwilym Ganes

Northern Ireland
Gary Cunningham, Ronan Coghlan

USA
Arkansas: Donnie Porter
California: Greg Bishop, Dianne Hamann
Illinois: Jessica Dardeen, Derek Grebner
Indiana: Elizabeth Clem
Michigan: Raven Meindel
Missouri: Kenn Thomas, Lanette Baker
New York State: Peter Robbins, Brian Gaugler
New Jersey: Brian Gaugler
North Carolina: Shane Lea, Micah Hanks
Ohio: Chris Kraska, Brian Parsons
Oklahoma: Melissa Miller
Oregon: Regan Lee
Texas: Naomi West, Richie West, Ken Gerhard, Nick Redfern
Wisconsin: Felinda Bullock

INTERNATIONAL
Australia: Mike Williams, Paul Cropper, Rebecca Lang, Tony Healey
Denmark: Lars Thomas
South America: Austin Whittall
Ireland: Tony 'Doc' Shiels, Mark Lingard
France: Francois de Sarre
Germany: Wolfgang Schmidt
New Zealand: Tony Lucas, Peter Hassall
Switzerland: Georges Massey
United Arab Emirates: Heather Mikhail

Dear Friends,

Welcome to another issue of *Animals & Men*, the third of 2015. Although some of my New Year's resolutions failed to come to pass, and some of them are still very much on the drawing board, my resolution to get this periodical back onto its original publication schedule of four issues a year, something which we have not achieved since 2003, is one that is going to plan.

There has been an unforeseen benefit to the new way of publication whereby the magazine is available free online as well as in hard copy and ebook formats. Previously we sold about 200 copies of the magazine, but when I came to look at the circulation of the new formatted version I had a big surprise.

A couple of days ago an old friend of us all wrote to me and apart from the usual social pleasantries wanted to know whether I was giving up cryptozoology, because I seemed to be spending so much more time doing stuff with and for Gonzo Multimedia these days. I was, I am glad to say, able to reassure him.

The CFZ, like everyone and everything else in the Western world, is a victim of governmental sponsored austerity measures. People who a few years ago were classed as unable to work, but who were able to do a few hours voluntary work for the CFZ here and there, have now been forced into low-paying menial jobs, and as well as facing serious health problems as a result of these new measures, are no longer able to do what they once could with the CFZ.

Money is, as *The Valentine Brothers* once so sagely pointed out, too tight to mention, and most of the quondam CFZ workforce have moved on as a result. The excuses that some have given (working towards the final months of a PhD, for example) are fine and admirable. Others, less so. Some have been completely ludicrous.

And it also has to be said that the media coverage given to cryptozoology, portraying it as a discipline way out on the lunatic fringe doesn't help.

The discussions on what were once - at least as far as I was aware - serious cryptozoological forums (I know that it should be *fora* but it just doesn't look right) about whether Bigfoot has access to 'cloaking devices', for example, is just one example of the ridiculous directions that many arms of cryptozoology are heading towards.

As the subject that I have championed for nearly half a century becomes the subject

The Great Days of Zoology are not done!

of ludicrous 'Reality TV' shows providing cheap entertainment to the gullible masses, it is not surprising that the people who *should* be working within the subject are leaving in droves.

But am I one of them? And is the CFZ on the way out?

The answer to both these questions is no!

Unfortunately, because my personal economics and those of the CFZ are not immune from the effects of the swingeing austerity measures of the British government, Corinna and I are spending more time than we used to do on other projects, with one sad exception - our monthly webTV show which is on indefinite hiatus - we are continuing to do what I always wanted to do: provide a clearing house for information, a daily web magazine, and a specialist publishing house for cryptozoological and - to a lesser extent Fortean - publications.

We do not publish as many non-cryptozoological Fortean books as we used to, but this is largely because we have lost interest in these subjects, and lost patience with many of the people who follow them.

I cannot foresee a time when we shall return to publishing a slew of books on UFOs, for example, but we are still publishing what I think is an impressive range of books on cryptozoology and allied disciplines, with new books by Dmitri Bayanov, George Eberhart, Michael Newton, Lee Walker and others either having come out in the last few weeks, or about to.

We publish a daily set of blogs on the CFZ Blog network mixing original news with digests of material on bigfoot, the Loch Ness Monster, and mystery cats, as well as general news and bird news of interest to cryptozoologists. We publish an annual peer reviewed journal which has largely taken the place of our old CFZ Yearbook (another project presently on hiatus, but which we intend to bring back at some point in the not too distant future) and we publish this magazine.

As I said at the beginning of this editorial, under the old regime, this magazine seldom sold more than 200 copies.

So I was expecting to have about the same level of readership for the new versions. Imagine my surprise, therefore when I found that - largely, I suspect, due to the good offices of Ronan Coghlan and Andrew May - at the time of writing, the last issue has had 15,004 readers.

And, for the first time in a decade we have an employee. Jessica Taylor has just left college and, a few weeks ago came to work properly for us. She has been doing odds and sods here on and off for years, but it is lovely to formally have her on board.

So no, I was able to write to our old friend. The CFZ may be changing, but reports of its death have been greatly exaggerated, and I truly believe that in many quarters we are still the only game in town.

Slainte

Jon Downes
(Editor, and Director CFZ)

A LEGAL MATTER

Wherever possible we use images that are either owned by us, public domain, or with the permission of the copyright holder. However, when we are unable to do this we believe that we are justified under the Fair Use legislation.

Copyright Law fact sheet P-09 : Understanding Fair Use
http://www.copyrightservice.co.uk/copyright/p09_fair_use
Issued: 5th July 2004

What is fair use?
In copyright law, there is a concept of fair use, also known as; free use, fair dealing, or fair practice. Fair use sets out certain actions that may be carried out, but would not normally be regarded as an infringement of the work. The idea behind this is that if copyright laws are too restrictive, it may stifle free speech, news reporting, or result in disproportionate penalties for inconsequential or accidental inclusion.

What does fair use allow?
Under fair use rules, it may be possible to use quotations or excerpts, where the work has been made available to the public, (i.e. published). Provided that: The use is deemed acceptable under the terms of fair dealing. That the quoted material is justified, and no more than is necessary is included. That the source of the quoted material is mentioned, along with the name of the author.

Typical free uses of work include:
Inclusion for the purpose of news reporting, incidental inclusion.

National laws typically allow limited private and educational use.

This magazine is not produced for profit, it is free to read and share. The hard copy version is available through Amazon and other outlets, but on a not for profit basis. We feel that we are justified in our use of copyrighted materials under several of the above clauses.

Newsfile

New & Rediscovered

There are Lost Worlds everywhere

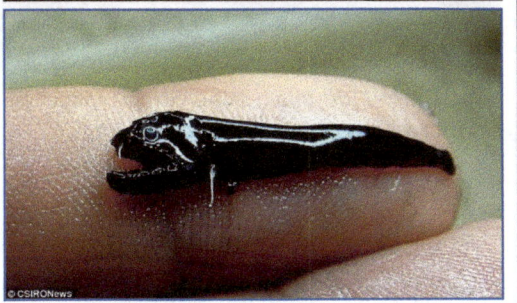

This tiny fanged 'blackfish' is among the new species discovered in a volcano off the coast of Australia. Scientists with CSIRO photographed the creature on a research voyage. They also discovered an eel-like creature classified as being one of the idiacanthidae and the fish pictured below from the chauliodontidae which boasts huge front teeth, Mail Online reports. However, the fish weren't the only thing found on the voyage, with four extinct volcanoes on the ocean-floor also discovered.

SOURCE: *Mail Online, 14th July 2015*

Not a New Toad for the UK

I am afraid that criticism of the poor reporting skills of the popular press looks as if it is going to be one of the *motifs* of this issue, and I promise that it is not because I am in a bad temper or a late middle-aged grumpy mood.

On 21st October 2014, (and I am mildly ashamed for having missed it) *Wildlife Extra* announced (and I quote):

> "Toads on the island of Jersey have been found to be an entirely different species to toads found in the rest of the UK.
>
> The Western Common Toad (*Bufo spinosus*) – which is also found in western France, Iberia and North Africa – is more genetically different from the Common Toad than humans are from gorillas or Chimpanzees. In the UK they are found only in Jersey, which is the only location in the Channel Islands to have toads."

Well I am, of course, pleased that there is a new species record from the Channel Islands, but they are not - either geographically or politically - part of the United Kingdom.

Geographically they are part of France where *Bufo spinosus* is a well-known species, and politically the Bailiwick of Jersey is a Crown Dependency. The Crown dependencies are the Isle of Man in the Irish Sea and the Bailiwicks of Jersey and Guernsey in the English Channel. Being three independently administered jurisdictions, they are not part of the United Kingdom or overseas territories of the United Kingdom. They are self-governing possessions of the Crown (defined uniquely in each jurisdiction). They are not member states of the Commonwealth of Nations, nor, save for a limited extent, a part of the European Union. However, they do have a relationship with the Commonwealth, EU, and other international organisations and are members of the British–Irish Council.

In the Channel Islands, the Queen is known informally as the "Duke of Normandy", notwithstanding the fact that she is a woman. The Channel Islands are the last remaining part of the former Duchy of Normandy to remain under the rule of the British monarch. Although the British monarchy apparently relinquished claims to continental Normandy and other French claims in 1259 (Treaty of Paris), the Channel Islands (except for Chausey under French sovereignty) remain Crown dependencies of the British Crown.

The British historian Ben Pimlott noted that while Queen Elizabeth II was on a visit to mainland Normandy in May 1967, French peasants began to doff their hats and shout "Vive la Duchesse!", to which the Queen supposedly replied "Well, I am The Duke of Normandy!" So, chaps. An island off the coast of France now has another French species living there, which - I am sure you will agree - is jolly good news.

SOURCE: http://www.wildlifeextra.com/go/news/jersey-toad-279.html#cr

The Man with the Golden Hutia

The *Daily Mail* much beloved newspaper of us all recently announced:

> "A new cat-sized rodent has been discovered on the Caribbean island Hispaniola.
>
> It weighs more than a kilogram, and has soft-brown fur and a short tail. It is also nocturnal and lives in small burrows or caves, which might explain why it has avoided detection for so long.
>
> Its name: Bond, James Bond. Or at least that's the version that rolls off the tongue. Its scientific classification is *Plagiodontia aedium bondi*."

The paper goes on to explain that the original James Bond, who gave both his name and mannerisms to author Ian Fleming, was a famous Ornithologist who lived and worked in the West Indies.

However, this is NOT a new species, rather a newly recognised subspecies of a well known (if extremely rare) species of rodent. As I have written in my book *The Island of Paradise*, the islands in the Antilles were home to an extraordinary degree of rodent speciation, and that many of the species are

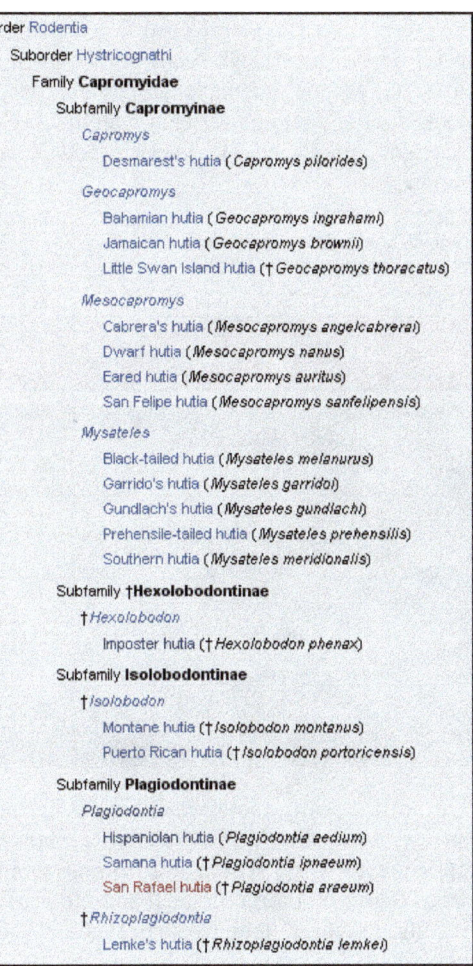

presumed to have become extinct in relatively recent times. As a result the rodents and the insectivores of the region are of particular interest to cryptozoologists.

So, all in all this is all rather good news. However, the galling thing about it is that the news is disseminated to the world wrapped in a whole slew of tabloid journalistic nonsense.

SOURCE: http://www.bbc.com/earth/story/20150515-meet-the-james-bond-rodent

The Blonde Bat

Having been long-mistaken for one of its relatives, a new bat species, *L. inexpectata*, has been now discovered. With their unusually pale fur, peculiar skull shape and tooth morphology, the specimens had spent long years in some of the most reputed nature museums behind the wrong sign, expert say.

A new species of nectar-feeding bat from Brazil was discovered unexpectedly amid a research into the whole genus of Lonchophylla. The study is available in the open-access journal *ZooKeys*.

During their study Drs. Ricardo Moratelli and Daniela Dias found that some of the specimens had their ventral (abdominal) fur considerably paler and some of their measurements were inconsistent with those of the type material of *L. mordax*, which species they had previously been confused with.

To their surprise, a closer look revealed that this was indeed a completely different species, previously unknown to science.

SOURCE: http://www.pensoft.net/news.php?n=516

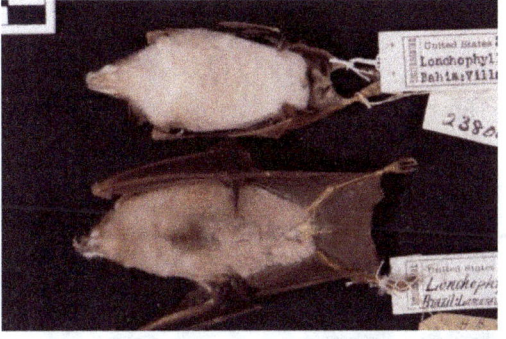

Shine on you Crazy Diamond

Meet the long-legged diamond frog, *Rhombophryne longicrus*, the newest species to increase the count of Madagascan amphibians once again. Like the rest of the diamond frogs, it is small and brown, but it is also very different.

Characterised by its unusually long slender legs, which are also the reason behind its name, the new species is unable to burrow its way through the ground like most of its relatives do. However, it makes it up with its longer leaps.

Unfortunately, the newly found diamond frog is likely at the risk of extinction. Lead-author Mark D. Scherz, a researcher at the Bavarian State Collection of Zoology, and his team preliminary assessed the *R. longicrus* frog to be Endangered in their present research, published in the open-access journal *Zoosystematics and Evolution*.

Little is known about the diamond frogs. However, one thing that has been found to apply to almost all of them is that they are burrowers. This is why they usually have short limbs, round bodies, and large, hard projections on their hands and feet, called 'tubercles' that help them dig. The long-legged frog described here, on the other hand, seemed unfit for this lifestyle, so the international research team looked closely into its skeleton with micro-CT scanning, producing 3D reconstructed X-ray models.

SOURCE: http://www.sciencedaily.com/releases/2015/07/150716124355.htm

Life in the Spray Zone

Up until recently there was a single known species in the only vertebrate family endemic to West Africa, the torrent toothfrog. Based on morphological and molecular results, however, four new species are now described. Unfortunately, they might all be at a risk of extinction. Their habitat needs and small distribution range call for immediate conservation measures.

The first species was only discovered last year. Now, Dr. Michael F. Barej from the Museum für Naturkunde, Berlin, and his colleagues verify the existence of as many as four new highly endangered species. In their study the researchers provide crucial insights

Another Compendium of Frogs

for the conservation of the biodiversity hotspot. Their research on the suggested existence of a complex of cryptic (structurally identical) species is published in the open-access journal *Zoosystematics and Evolution*.

SOURCE: http://www.sciencedaily.com/releases/2015/07/150727120218.htm

Big Headed Robber

Herpetologists with the Bolivian Faunal Collection and the National Natural History Museum have discovered a new species of robber frog (*Oreobates sp. nov.*) in Bolivia's Madidi National Park. The frog, also known as the big-headed frog, is from the Craugastoridae family of amphibians and was found during the first leg of an 18-month herping expedition to chronicle the wildlife in the park. The robber frog is a small to medium-sized frog that can be found in the Andes and Amazon region of Bolivia and surrounding countries. There are 23 known species in the family that this frog belongs to, according to James Aparicio, who, along with fellow herpetologist Mauricio Ocampo, discovered the species.

SOURCE: http://www.reptilesmagazine.com/Frogs-Amphibians/Information-News/New-Robber-Frog-Discovered-in-Bolivias-Madidi-National-Park-Trending/

Sounds like a Blues Singer

Carrying itself around with a dark brown mask on its face and a broad shapeless white mark on its chest and belly, a frog had been jumping across the Peruvian cloud forests of the Andes unrecognised by the scientific world.

Now, this visibly distinguishable species has been picked up by Dr. Catenazzi of Southern Illinois University and his team from its likely only locality, a cloud forest near Cusco in Peru, at 2350 m elevation by Drs. Catenazzi, Uscapi and May.

Their research is published in the open-access journal ZooKeys.

SOURCE: http://www.sciencedaily.com/releases/2015/08/150806112051.htm

Another Compendium of Frogs

Bat with a Long Tongue

A groundbreaking Bolivian scientific expedition, Identidad Madidi, has found a bizarre bat in Madidi National Park. The researchers found the bizarre tube-lipped nectar bat (*Anoura fistulata*) -- the first record of this species in the park. Described in Ecuador just a decade ago and known from only three records. It has the longest tongue in relation to its size of any mammal -- stretching 8.5 cm to reach into the deepest flowers.

SOURCE: http://www.sciencedaily.com/releases/2015/08/150821093138.htm

Fly Away Home

A fly thought to be extinct in the UK has been found in a Devon nature reserve. The *Rhaphium pectinatum* was last recorded in Britain 147 years ago in 1868 but was rediscovered in Old Sludge Beds on the outskirts of Exeter.

The fly is from the Dolichopidiae family, a group known as long-legged flies, and is usually found in tropical parts of the world. Devon Fly Group member Rob Wolton said he was surprised by the find. The last recorded sighting was on 19 July 1868 when the Victorian entomologist George Verrall caught a male and female at Richmond in south-west London.

SOURCE: http://www.bbc.co.uk/news/uk-england-devon-33707788

Grande Cojones

A recently discovered lemur from Madagascar has the largest testes per body weight of any primate, new research finds. If the northern giant mouse lemur were the size of a human, its testes would be as big as grapefruits, said Christoph Schwitzer, the director of conservation at the Bristol Zoological Society in the U.K.

SOURCE: http://www.livescience.com/51563-lemur-has-biggest-testes.html

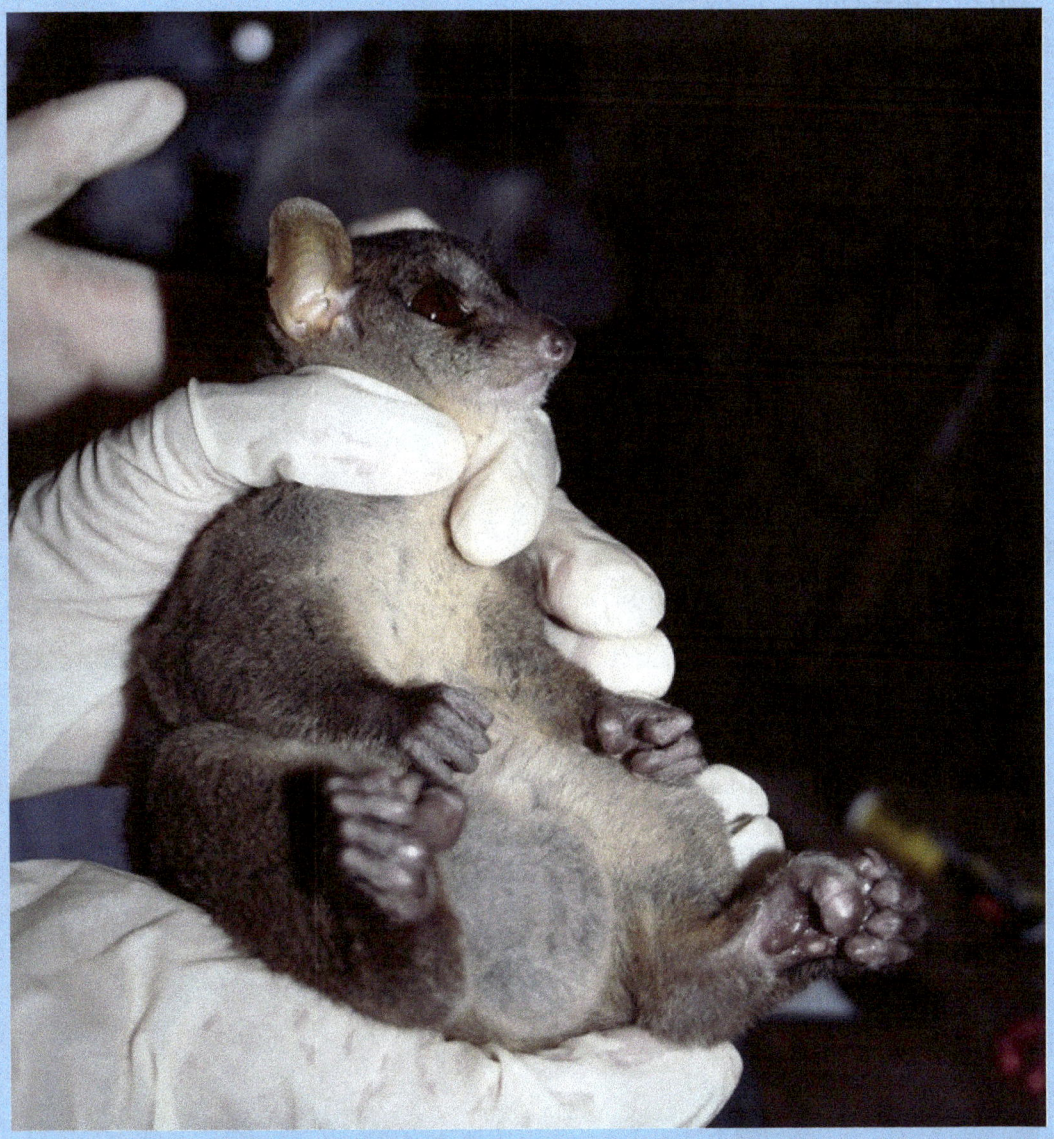

Goodness me, they *are* big. So much so that they remind me of a certain song by the late Peter Cook. Mouse lemurs are also known for their sperm competition. During breeding seasons, the testicles of male mouse lemurs increases in size about 30% of their normal size. This was speculated to increase the sperm production thereby conferring an advantage for the individual to bear more offspring. There are various hypotheses relating the rapid evolution of mouse lemur species this sperm competition. Mouse lemurs are considered cryptic species - with very little morphological differences between the various species, but with high genetic diversity. Recent evidence points to differences in their mating calls, which is very diverse. Since the mouse lemurs are nocturnal, they might not have evolved to look differently, but had evolved various auditory and vocal systems.

Chupacabras

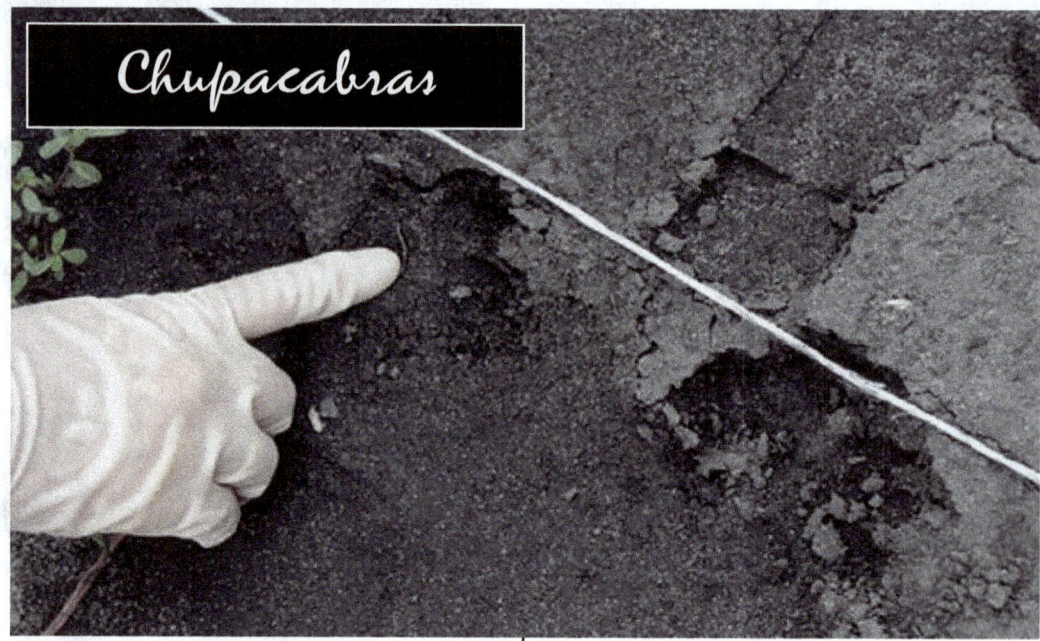

I noted same years ago that the term *chupacabras* literally meaning the sucker of goats is no longer just used to describe the semi bipedal spiny creature that has been reported for the past twenty years or so in and around the upland Canovenas plateau on the island of Puerto Rico. The term has spread across the Hispanic world, and now is used to describe *any* mysterious creature, usually one suspected of vampiric tendencies.

In recent years it has also been used to describe

the hairless blue dogs which have been reported across the United States, but not the term seems to have spread to Russia. A series of mysterious killings began in July 23rd in the village of Davydovka, which is about 300 miles from Moscow. One family found 95 mutilated chickens. Two other families lost a combined 100 birds the following day and another lost an unspecified number of chickens and 60 geese a few days later.

According to reports, all of the farm families said the chickens had bite marks but no or very little blood on the wounds or the ground. No one reported hearing any sounds of startled chickens or a struggle. What killed the chickens of Davydovka? The usual suspects are dogs or a wolf. However, the only set of tracks were large enough to be from a 40 kg (88 pound) creature – much bigger than any of the dogs in the area.

SOURCE: http://mysteriousuniverse.org/2015/08/chupacabra-blamed-for-mysterious-chicken-deaths-in-russia/

There have been a whole string of so called "Chupacabras" reports in recent months, and they have all – or at least the ones I have seen – been of ill looking canids suffering from a range of skin diseases.

Whilst I truly believe that a small number of these painfully ill creatures are truly of scientific importance because of their singular behaviour the way that they appear to breed true, and the inexplicable fleshy pads on their haunches, it is unarguable that most of the animals reported to the CFZ newsdesk are nothing more dogs, coyotes, and occasionally other animals with mange.

I refer all interested parties to the lecture given by my step-daughter Shoshannah McCarthy at the 2014 Weird Weekend; it can be seen on YouTube.

Meanwhile our investigations, mostly via Richie and Naomi West in Killeen, Texas, continue.

Blue Dogs

Man Beasts (BHM)

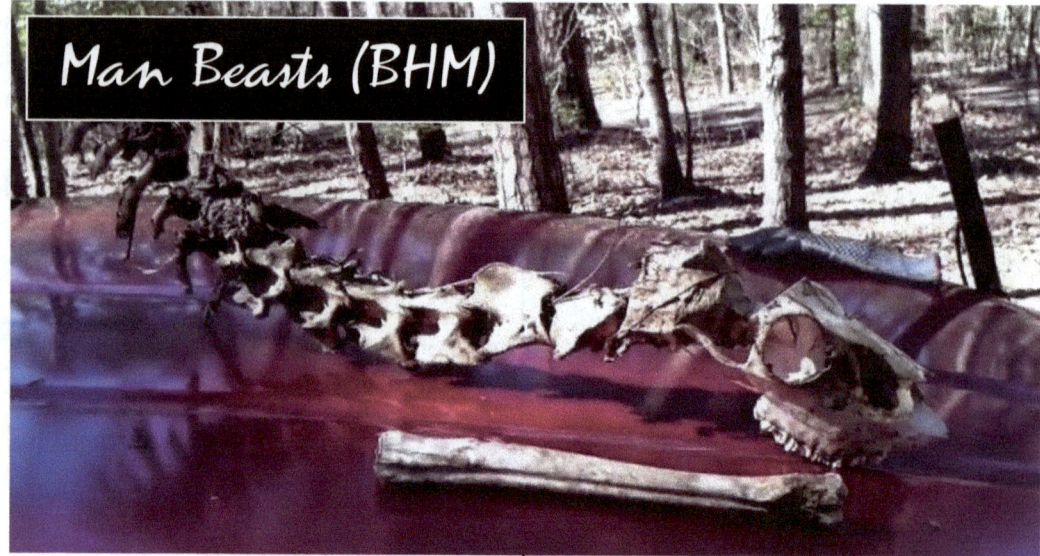

There have been a number of what are purported to be the bones of a bigfoot-type creature in recent months.

In March, for example, the photograph above was posted on the *Bigfoot Evidence* blog, with a caption:

"Researchers in North Carolina made a grizzly discovery of some type of skeleton buried beneath a pile of sticks and debris. They're having a hard time identifying what kind of animal it came from. Can you identify it?"

The second story, which accompanies the pictures on the next two pages is a little more macabre. The story on the *Sasquatch Chronicles* blog reads:

Did anyone ever hear anymore about this? Before Chuck Prahl passed away he wrote "I was contacted by a man in Oregon named Craig who had these pictures sent to his phone by his former brother in law who was out hunting with a friend in Grants Pass along the Illinois river a few years back when the discovered these bones near the river.

The two hunters were amazed by the size of the feet and length of the leg bones. They thought at the time they might be a big human or maybe even bigfoot. They called the local game wardens who came up to the inspect the find. The hunters said that within a hour or so 2 black suburban pulled up and these men confiscated the bones and told everyone there they never saw this "It didn't Happen" Just for get about it. They also confiscated their cell phones but didn't know they had already sent the pictures to Craig and other family members.

One of the posted comments refers to the second picture:

Obviously human bones, you can see the top part of a sock on the last one!

I would have to say that it does look definitely sock-like!

There seemed to have been a spate of sightings of a Bigfoot-like creature in the revolutionarily named Provo Canyon in Utah, with a rash of news stories accompanying a picture of a vaguely ape like shape partially obscured by foliage, which appeared on lots of websites during August 2015.

> The story which accompanied this picture read: "We went camping in Provo Canyon (near Squaw Peak and Little Rock Canyon Overlook) and saw some deer up on a hill that we wanted to get a closer look at.
>
> On our way up, we thought we saw a bear, until the monster stood up and looked right at us. We ran straight to the car after that, leaving our tent and everything behind. It's probably all still up there

However, because of my Hibernian connections the name had amused me, so it had stuck in my mind, and I remembered seeing reports of a similar sighting there a couple of years ago. I had indeed. But it was not a "similar" sighting at all. It was exactly the same sighting. A cursory investigation online proved that this fairly unlikely story seems to have surfaced first towards the end of 2012. It surfaced again in the summer of 2014, and again this year. The un-named protagonists thought that what they had seen was a bear, and I tend to agree with my colleague, the lovely Sharon Hill at Doubtful News, that a bear is exactly what it was.

Mystery Cats

What made Milwaukee Famous

I don't know whether there is a branch of sociology that deals with the way that the mass media promulgates and proliferates news stories, but if there isn't there jolly well should be. Not for the first time this issue I am feeling mildly guilty that I seem to have spent more time slagging off the media than anything else this issue, but it really has been an extraordinarily interesting time recently for crypto–media watchers. Take this story, of the Milwaukee lion, for example. This story started in mid July when Fox6 Now reported.

"Video surfaced Tuesday showing what appears to be a big cat sauntering through a neighborhood. The woman who shot the video planned to grill out on Tuesday evening — but with a baseball bat nearby — just in case."

The video is very indistinct and Jessica Taylor who works in the CFZ office was not the only person who wondered why, in the 21st century when everyone has mobile phones capable of taking clear and distinct images, the video is so blurry.

The event had allegedly taken place on 20th July, and two days later the couple who had taken the film were interviewed by www.gngn.com:

"Bill Nolen and his wife, Annie, who are both residents in Brewers Hill, took a cell phone video of the big cat walking around the neighborhood. But despite the video, no big cat was spotted in the area.

"I was afraid to move," Annie Nolen said, WISN-TV reported. "I was sitting there, and I couldn't move. I thought, 'What am I looking at?'" When the police received a call about the sighting, they

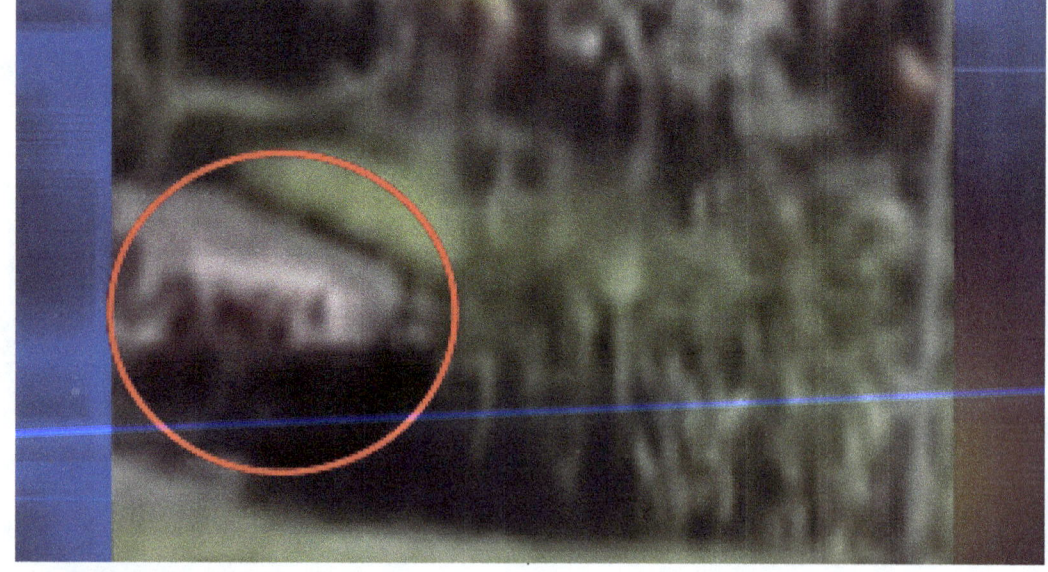

checked the area for several hours but didn't find the animal."

Then the police got involved and various news reports claimed that they found the reports "credible". On the 26th *Reuters* announced:

"Police closed off streets near a park in north Milwaukee on Saturday night after reports of a "lion-like animal," which came days after people said they had a seen a lion prowling city streets. Milwaukee police and the Department of Natural Resources responded to the area, at 30th and Fairmont streets, for what a police statement called "a confirmed sighting of a lion-like animal."

The animal was not located, it added, but police would maintain a presence throughout the night. Local residents reported earlier on Saturday they had seen a lion and her cub in a creek in the park."

Then a few days later Fox News announced:

"Milwaukee police on Tuesday night, July 28th responded to 61st and Fairmount for an "unconfirmed" sighting of the possible Milwaukee lion. Nothing was found. A Milwaukee Police Department spokesman said earlier Tuesday the department took calls Monday and Tuesday regarding a possible lion in Milwaukee. Officers have investigated each and every call — and so far, nothing has been found.

"Milwaukee police remain committed to protecting the safety of the public and our officers. To that end, MPD has received advice from a variety of animal experts, including those who specialize in large cats. The Milwaukee Area Domestic Animal Control Commission (MADACC) and the Wisconsin Department of Natural Resources have dedicated resources and are providing assistance," the MPD spokesman said in a statement."

One of the things that I have written widely about during my career as ringmaster of the world's largest cryptozoological circus is something I have dubbed 'The Mythologisation Process'. This is the socio-cultural process by which myths are made.

One classic example can be found in the way that the media deals with the predations of a small population of naturalised big cats which appear to live on the moorlands of southwestern England.

If there was a learned article entitled something like *Feeding Patterns of a*

Animals & Men   Issue 54

group of naturalised *P.concolor* on Westcountry moorlands no-one would take any notice, but when *The Daily Mirror* proclaims *The Beast of Bodmin Strikes Again* then it sells a lot of newspapers.

Within a week of the initial sightings the Mythologisation Process had kicked in and not only were people smelling lions where there are no lions, if I may coin an Ahabism, but it had become an internet meme with its own Facebook page and a selection of photoshopped pictures depicting a lion in incongruous locations across the city.

In the meantime I took a closer look at the original picture, and my 86-year-old mother-in-law pointed out that the shadows underneath the creature made it look very much as if the creature was walking along the top of the stone wall. This would make the creature a somewhat muscled domestic cat, and the wall a low one such as the one around someone's suburban rockery. But what do we know?

Aquatic Monsters

It ain't Nessie Cerally so

In mid July Steve Feltham, possibly the most well-known contemporary Loch Ness Monsters researcher, announced that in his opinion the animal from which he has been seeking for the past two decades or more is nothing more than a giant Wels catfish. This European species regularly reaches a length of 6 feet and has been known to reach 8 or 9. Even longer specimens have been claimed, but these records are both unreliable and ancient. They were certainly introduced into the United Kingdom by The Acclimatisation Society in the late 19th century, and the latest addition of Sir Christopher Lever's *The Naturalized Animals of Britain and Ireland* pin points nearly thirty locations where these fish have been found.

In my book *Monster of the Mere* (2002) I

describe the events which led us to conclude that there is a particularly large specimen of this species in Martin Mere in Lancashire where it was seen by a number of witnesses including out own Richard Freeman.

I only met him once, but I have the greatest respect for Steve Feltham, and although I do not agree with his hypothesis, I both understand and respect his reasons for formulating it. What I find most disturbing is the way that the national press and many internet pundits have treated this news story. It has been claimed that:

- There is now conclusive proof that the Loch Ness monster doesn't exist
- That the Loch Ness monster is dead and that Steve is leaving the lake for good
- The mystery has been solved once for all

All of this is, of course, arrant nonsense!

Steve is not leaving the lake, nothing has been proved, and there is no proof whatsoever that anything has died. Steve has merely promulgated a perfectly cogent hypothesis and a bunch of idiot journalists have jumped to a bunch of airy fairy conclusions. We at the CFZ wish Steve's future efforts at the lake all the best and look forward to reading his reports from the lochside.

EDITOR'S NOTE: For the record Richard Freeman and I have always tended towards the Eunuch Eel hypothesis, and no cryptozoologist worth his salt has ever considered the idea that there is some kind of *Jurassic world* in the Highlands of Scotland.

It ain't Nessie Cerally so

The Metro in February 2015 described a mysterious animal seen in the historic waters of Plymouth Sound:

"A large unrecognisable object was spotted off the coast of Britain – so naturally people have started thinking a crocodile or some sort of monster is lurking in our waters.

Photographer Allan Jones took some pictures of the creature wallowing in the waters off Plymouth Sound in Devon.

He said the '20 foot long' animal was swimming against the current about half a mile offshore, and when he showed the images to a university technician he was told it was probably a crocodile.

Allan said: 'I've never seen anything like it – the first thing that struck me was that it looked just like a huge crocodile. 'The creature or object moved in circles, appeared to curve its shape and moved a considerable distance from left to right, turn and then move back the other way."

Well, first of all at the risk of being a complete spoilsport, although the Indopacific crocodile makes immense sea journeys, and has even turned up as far north as Hong Kong and as far south as New Zealand, it is not a species likely to be found in the waters off South Devon.

My personal guess is that it is a rather manky basking shark, but if there is anyone out there who disagrees with me, please get in touch via the editorial address.

The Devil and the Deep Blue Sea

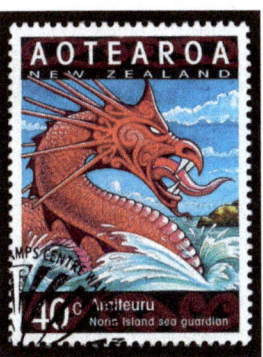

New Zealander Pita Witehira was looking at Google Earth photos of the area around Oke Bay, where he has property. He saw what he interpreted as the wake of a creature that appears to be 12 metres (39 feet) long. There have been a number of media furores in recent years when people have interpreted boat wakes on Loch Ness and Loch Lomond (which has a far less impressive history of monster sightings) but this is the first time, to my knowledge, that something of the sort has been claimed from the open sea.

Admittedly the image here doesn't really look like a boat wake, but to these eyes, neither does it look like a monster. Mysterious Universe on December 16th 2014 quoted Witehira as saying:

"It's got to have a lot of weight under the water to create that kind of drag. The native Maori would call this a Taniwha as it appears not to be a whale and it is far too big to be a shark. It is moving too fast and turning too sharply to be a whale".

Mysterious Universe added:

"The Maori believe the Taniwha are supernatural creatures that live in New Zealand rivers, lakes and coastal waters. They can be shaped like sea serpents or winged dragons or giant sharks and may even be shape-shifters. They are described as guardians of tribes and of the waters but many legends tell of them killing and eating people and kidnapping women. The belief is still strong in modern times as a highway was redirected in 2002 because the native people feared it would destroy the home of their Taniwha."

Sadly I remain unconvinced.

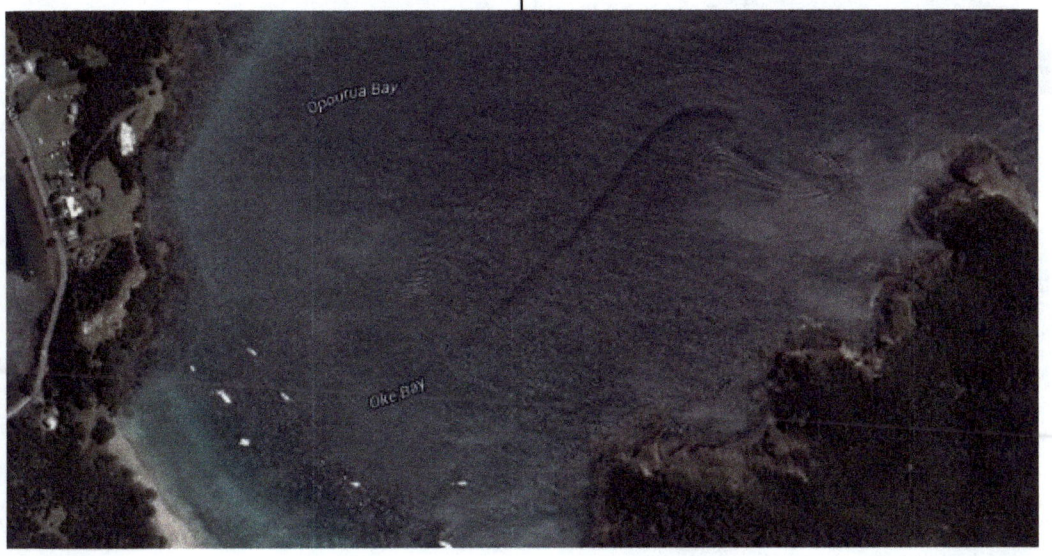

Carl Marshall's Column

The Alcester Seal
The River Arrow in Alcester, Warwickshire, became the focus of attention during last January - March [2015] after bizarre reports were sent to the *Wildlife Trusts* of alleged sightings of a grey seal *Halichoerus grypus* feeding on fish in the river. My first thought was that these reports must be down to misidentified observations of European otters *Lutra lutra* which have of course made a comeback in recent years and can now be found countrywide, since completely vanishing from every county except the West Country and Northern England. An otter swimming could perhaps be mistaken for a seal if the observer had not personally seen either of these species before in the wild.

Apparently, in November 2014, Alcester resident, Claire Romos noticed a *"very heavy trail through the grass on the river bank, with scales/bones, flesh, evidence of something large taking fish"*. This description could again be attributed to the European otter, however luckily a photograph was taken by Claire, which can be viewed at *www.warwickshirewildlifetrust.org.uk.*

The photograph is a little blurred but is certainly that of a grey seal! It is copyrighted by Claire Ramos and is the original photograph she took at the time and therefore not a stock image.

So if there really was a grey seal living in the River Arrow at Alcester how did it get there? Maybe it was a released pet? Perhaps it teleported to Alcester from Charles Fort's Super Sargasso Sea!! Or could it have travelled upstream from the River Severn? I was pondering upon the possibilities when I remembered the saga of Keith the Seal!

Keith The Severn Seal
By coincidence, after the publication of *Worcestershire Mammal Atlas* (Green *et al*

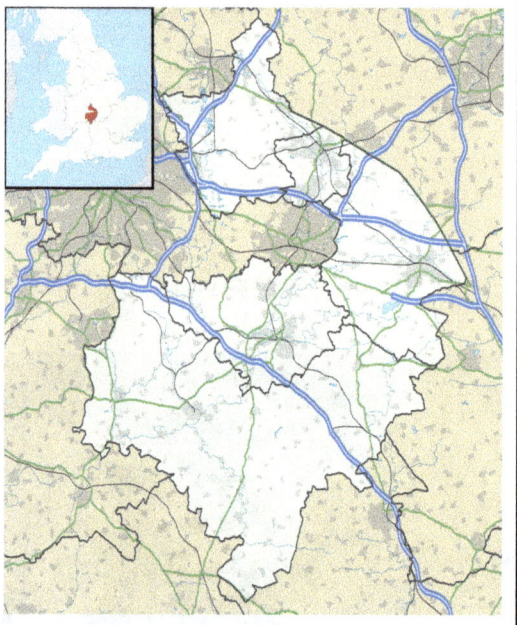

The Alcester Seal and Other Unexpected Aquatic Guests

2012) with no reference to the species, a grey seal was sighted in the county along the River Severn. This provided headline news for local papers and sparked discussions for naturalists and anglers. Her inquisitive nature endeared her to many onlookers and she became quite the celebrity. This surprising visitor brought people and nature together in a unique way and Keith the Seal even had an internet Facebook page and could be followed on Twitter.

Members of the *Wychavon Kayak and Canoeists Club* have the unique claim of first sighting Keith (gender explained below); she was spotted on the 17th November 2012, as the canoeists were enjoying their sport on the River Severn near Powick, south of Worcester. She was observed for approximately half an hour by the kayakers while she swam about between their boats apparently enjoying human contact, and they were able to take several photographs and a video of their encounter with her. It was the kayakers who first named the seal Keith after the Scottish aristocrat Royalist Commander Colonel George Keith, who fought for King Charles the 1st against the Roundheads during the English Civil War and whose troops were eventually defeated as they tried to defend Powick Bridge in the Battle of Worcester in 1651.

The Worcester seal was soon confirmed as a female, but the name Keith prevailed for the duration of her stay. Over the next few weeks she was photographed at a number of locations along the River Severn.

The details below denotes dates, locations and behaviour attributed to Keith.

Date	Place	Activity
Nov 17 2012.	Near Powick.	Eating salmon.
Nov 18 2012.	Diglis Weir.	Eating large chub and 2 pike in 30 mins.
Nov 22 2012.	Ketch Caravan Park.	Playing with plastic bottle.
Nov 30 2012.	Upton on Severn.	
Jan 1 2012.	Bewdley Severnside North.	Eating fish.
Jan 2 2012.	Stourport Marina.	Present all day eating fish.
Jan 4 2012.	Bewdley.	In river close to the town.
Jan 5 2012.	Bewdley.	Playing around canoeists.
Jan 6 2012.	Near Dowles.	Caught fish then came to rest in the shallows for 2 hours.
Jan 12 2012.	Bewdley.	
Jan 13 2012.	Near Dowles outflow.	With rowers, close to many people, eating mallards.
Jan 17 2012.	Bewdley.	Eating fish.
Jan 19 2012.	Lenchford Hotel.	On the bank in the snow.
March 9 2012.	Below Gloucester.	With paddle boarders, tried to climb on board.
May 14 2014.	Kempsey.	Emerging from the river.

There have also been a few seal sightings thought to be Keith reported from Upton Marina at Brockweir on the River Wye, though these could actually be of a different grey seal as reports differ concerning markings on the Brockweir animal's skin.

Keith spent over two weeks in Bewdley, Worcestershire, during January 2013 and then disappeared, not being spotted again until May 2014 at Kempsey, then disappearing again.

To swim from the sea, up the Bristol Channel as far as Bewdley, Keith must have encountered many weirs. Before these weirs were constructed, seals and porpoises were occasionally reported in Worcestershire's rivers and even Atlantic sturgeon *Acipenser sturio* were once recorded as far upstream as Worcester.

[It is believed that the sturgeon specimen on display at the *Worcester City Museum and Art Gallery* was captured in the Severn at Worcester on 25th July 1835. At the Autumn meeting of the *Worcestershire Natural History Society* in 1835, Sir Charles Hastings reported that *"the specimen of sturgeon caught in Worcester this year had been presented to the museum"*. This being the case, it would make this one of the earliest acquisitions to the museum's collections, coming just two years after the museum was founded in 1833. At one time, it was believed that it could be the fish that was caught in the River at Diglis in 1843; this fish at the time was placed against the south wall of the water gate and its shape cut out around it with a chisel - which incidentally can still be seen today. However, after measuring both, it is clearly not the same fish, as the sturgeon kept on display at the top of the museum's main

staircase is six feet (61cms) longer.

Even in medieval times the value of sturgeon was appreciated. In 1324 the King of England, Edward II, decreed the sturgeon as a 'Royal Fish', by an Act of Parliament which, in theory still stands today and claims any fish caught in British fresh waters belongs to the sovereign. Although the fish recorded in our waters are normally stragglers coming from Scandinavia and the Black Sea area, they were not as unusual as they would be today. This is again because there are more weirs, locks and other water features that would prevent them from reaching water as far upstream as the midlands (River Severn and Wye). This can also be confirmed throughout Europe, as sturgeons were already recorded on the Shannon, Rhine and Danube, these are now rare occurrences. Pollution is also said to have taken its toll on the fish, as the number of breeding adults has dwindled considerably.

Other captures from Worcestershire include several sturgeons recorded in 1849, captured by Mr W. V. Ellis - one of these fish was six feet six inches in length and 200lbs in weight. Another sturgeon measuring six feet one inch was captured in 1869 by a local fisherman named George Jenkins, while netting for salmon in the vicinity of Diglis weir.

On Wednesday (date unknown) two fishermen, on drawing their net in the Severn close to Diglis lock, found, to their surprise that they had captured one of those rare visitors - a young sturgeon of about four and a half feet. On the same morning another sturgeon, eight feet in length, was caught in the Severn near Tewkesbury and exhibited alive at the local Cross Keys Inn.

A 'Fine Sturgeon' weighing 11.2 cwt, and measuring seven feet five inches, was caught in the River Severn at the Teme's mouth in July 1872, by John Jenkins a local fisherman. The royal fish was sent the same day to her Majesty at Windsor Castle, by Mr James a fishmonger from Broad Street. This was the second large sturgeon caught in the Severn in 1872!

Sturgeons are fish of the North Atlantic Ocean, only coming into freshwater to breed; it is here where they are most likely to be seen. They can reach a size of twelve feet (365cm) or more in length, making them the largest species of fish that would normally be found in British freshwater!

Pinnipeds revisited
An act of Parliament was passed in 1842 to allow the River Severn to become navigable between Gloucester and Stourport. During the next thirty years, not only were six locks and weirs constructed, but the river was dredged and deepened by the removal of rock bars, so that a consistent depth of ten feet was obtained for the passing boats.

How could Keith have managed to find her way so far upstream beyond these weirs? The lowest weir at Maisemore, near Gloucester, is deluged by each high tide and seals have been recorded in the river as high as Tewkesbury weir fairly frequently, but records above this are very unusual. The next weir near Tewkesbury, the upper lode, is covered by exceptional high tides, but all the other weirs upstream are only saturated when the River Severn is in flood. Records show there were flood conditions on the Severn at various times during the late autumn and winters of 2012/2013, providing an opportunistic seal the chance to cross the weirs with minimal exertion as she followed the salmon coming upriver to spawn.

Could this be another seal or could Keith have found her way as far as Alcester?
Soon after March 2013, sightings of Keith stopped, leaving residents hoping she had made her way back downstream, suggesting human intervention would be unnecessary. One late record and an internet video clip of her near Gloucester playing with some paddle boarders on the 9th March, and a late encounter by a family on 14th May 2014 seems to be the last confirmed sightings. That is until late January 2015, when the *Wildlife Trusts* started receiving the reports of the Alcester Seal!

Its unusual for grey seals to choose to live in freshwater, but not unheard of. If this was Keith in the River Arrow, and she had not yet made her way back to sea or been illegally destroyed, there must be something about this environment she liked. Maybe she was here due to a shortage of resources off the Atlantic coast! Whatever the reason, its important to monitor this movement to keep a check on protected species of fish that also call the river home. One or two adventurous seals like Keith probably will not have much of an impact on river ecosystems as a whole, but if lots more choose a life on the river, it might represent a significant threat to several rare and protected fish species, not to mention the impact the presence of seals would have on renewed otter populations!

On a lighter note though, while in Worcestershire, Keith provided the public with a compulsion to follow her on her unique journey, and she brought people and wildlife together in a very special way.

A special thank you to *Worcester City Museum and Art Gallery* for allowing me to view and photograph the Sir Charles Hastings sturgeon.

References and suggested further reading:

- **Beach, G. (2015).** *Wild Warwickshire, A Warwickshire First? Grey Seal in Alcester.* Warwickshire Wildlife Trust. UK.
- **Burtzev, L.A. (1999).** *The History of Global Sturgeon Aquaculture.* Journal of Applied Ichthyology.
- **Drummond, R & S.** *Canoeists guide to the River Severn.*
- **Kruuk, H. (2006).** *Otters: Ecology, Behavior and Conservation.* Oxford University Press, NY. USA.
- **Reporter, Daily Mail. (2013).** *Keith the Seal Saved!* The Daily Mail Online. www.dailymail.co.uk
- **Sargent, G. & Morris, P. (1999).** *How to find and identify mammals.* The Mammal Society. UK.
- **Unknown, (2015).** *Save Keith.* WWW.FACEBOOK.COM/SAVETHESEAL.
- **Wikipedia,** *the free encyclopedia.* https://en.wikipedia.org.

> Carl Marshall works at Stratford Butterfly Farm and is a fine field naturalist. Over the past couple of years he has become a very enthusiastic member of the CFZ, and his quasi-Fortean view of British natural history fits in perfectly with my own. He was, therefore, the perfect choice as a columnist for the brave new *Animals & Men*, and we are proud to have him aboard.

Carl Portman's Fortean Invertebrates

Guilty as charged M'lud. This time around I have left my invertebrate comfort zone and stomped headlong into the world of…bats. I am so happy that my loft is 'protected'. Not with a security alarm, but the fact that it houses long eared bats.

They roost here every year which gives me and my wife Susan the greatest pleasure. Not only do they keep the mosquito population down around the homestead, but their presence indicates that the immediate environment around my house is a good one – otherwise they simply would not roost here.

I noticed them first when up in the loft a few years ago I was looking for something when the wall began to move – I shone my torch and reflected many pairs of little beady eyes staring back at me. I don't know who was more startled, me or the bats. I took a very quick photograph of three of them huddled near my head, (figure 1) then left them alone for the summer.

I should have known we had them though as a couple of weeks previously my slumbering wife was unaware of the gorgeous little bat flying around her head in the bedroom! I managed to catch it (don't ask) and put the tired little fellow outside.

Bats are seriously misunderstood of course. Some people believe them to be birds, others think that they are flying mice (Mozart?) and others have even ventured that they are reptiles, but we of course know that they are warm blooded mammals. They are utterly curious creatures, put together perhaps by the same 'manufacturer' of giraffes, camels and the platypus. I have enjoyed learning more about bats in recent years and I should like to share a little with you.

Figure 1 *Long eared bats at Portman Towers*

You might be surprised to learn that there are 16 breeding species of bats living in Britain. There were 17 but the greater mouse eared has disappeared it seems. My long eared bats are (I think) brown long eared, not the grey long-eared. They are curious creatures and much feared by some. Of course the old wives' tale about them deliberately flying into your hair is nonsense, but that has never stopped the hype. Bats are used in a negative way in films such as *Dracula* of course. The vampire bat sends shivers down the ignorant spine, but of the 1,100 species of bat on the planet only three species feed on blood. Let us just have a look at the morphology of a bat. *(Figure 2)* before I move on. You can see that it is a creature that has thumbs, elbows and wrists!

I think this is just the publication to discuss **myth and folklore** so let us do just that.

Bat Myth and Mythology
In Tonga and ancient Babylonia bats were considered physical manifestations of the Souls of the Dead. In China and Poland they were symbols of Happiness and Long life, and to the ancient Mayans they symbolized Transformation and Rebirth.

Perhaps because of their nocturnal habits and ability to navigate in the dark, or simply because they appear to be both animal and bird at the same time, bats have long been associated with deity, supernatural forces and the occult. In the mythologies of differing cultures bats symbolize both good and evil, life and death. In China many legends associate bats with good fortune. To them, a group of five bats represented the five causes of happiness: wealth, health, long life, virtue and a natural death.

In South America among an ancient Mayan cult of the Quiche, located in the jungles of what is now Guatemala, Camazotz was a minor deity associated with bats. He was the God of the Caves and is described as having the body of a human with the head and wings of a bat. According to the Mayan sacred book of initiation rites Popul Vuh, he resided in the Bat-house located in the Underworld, a

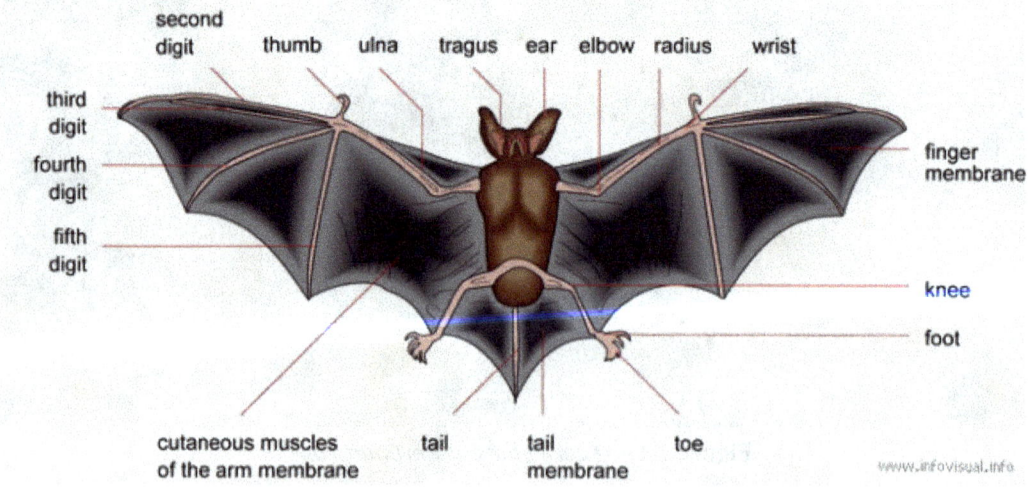

MORPHOLOGY OF A BAT

labyrinth of caves through which huge bats flew. While legends differ, he was responsible for the seventh test of initiation undertaken by the Mayan Hero Twins, the mythical Mayan ruling deities.

In ancient Greece and Rome, it was thought that sleep could be prevented either by placing the engraved figure of a bat under the pillow, or by tying the head of a bat in a black bag and keeping it near to the left arm. On the Ivory Coast, even today, many think that bats are the spirits of the dead, and in Madagascar, they are assumed to be the souls of criminals, sorcerers and the unburied dead. In medieval Europe, bats were commonly thought to be witches' familiars. In France 1332, Lady Jacaume of Bayonne was publicly burned simply because bats were seen to fly about her house and garden.

Also in Europe, in the Tyrol regions of Austria, it was believed that if a man wears the left eye of a bat on his person, he may become invisible, and in areas of central Germany, if he wears the heart of a bat bound to his arm with red thread, he will always be lucky at cards. It was commonly thought that witches used the blood of bats as an ingredient when making flying ointment, and further, to boost the powers of their magical brews and potions. To the Gypsies, who were equally ostracized as witches, bats were seen as the bearers of good luck; they even prepared small bags containing dead bat bones for children to wear around their necks as charms.

In folklore, to wash your face in bat's blood will enable you to see more clearly in the dark. To keep a piece of bat bone in your pocket will ensure good luck. Powdered bat's heart will staunch bleeding or stop a bullet, and bullets from a gun swabbed with a bat's heart will always hit their target. To put bat's blood into someone's drink will make him or her more passionate, and you can stimulate a woman's desire by placing a drop of bat blood under her pillow. To prevent baldness or your hair greying, you should wash the hair in a concoction of powdered bat wings and coconut oil. The list of folklore concerning bats is endless, and even Shakespeare got in on the act. In his famous play *Macbeth*, he had his three witches adding "wool of bat" to their hellbroth, and in *The Tempest* (Act I, Scene 2) he had Caliban place a curse on his master Prospero, which included the line: *"All the charms of Sycorax, toads, beetles, bats, light on you!"*

Bat as a totem animal
The bat as a totem animal is a symbol representative of transition and rebirth. A bat appearing in your life could mean that some aspect of your life is coming to an end, and rather than fear the change, you should embrace the transition and look forward to some kind of new beginning. It's a time for serious self-examination and self-evaluation. This may sound easy to do, but for most people change is a frightening experience. Bat's appearance is there to help you soar above your fears by getting rid of those things in your life that are no longer needed. Only by facing the darkness of your uncertainties can you progress and find light in new beginnings.

To many misinformed people, the bat is a symbol of death, but try to embrace the positive powers of the bat. Bats typically live in deep underground caves, which symbolically is the belly of the Mother (Earth), and from these womb-like caves they emerge each evening at dusk - reborn. To a shaman the appearance of a bat does not signify actual or physical death, but more the death of old fears or the old ways of doing things that no longer serve you. By learning from the bat you can fly through any darkness into the light, be transformed, reborn and free.

"For as the eyes of bats are to the blaze of day, so is the reason in our soul to the things which are by nature most evident of all" - Aristotle (384-322 BC).

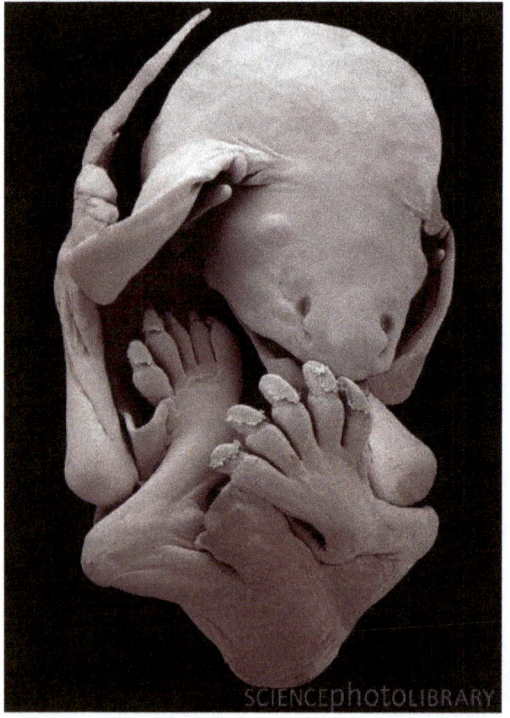

Bats look curious even before they are born Take a look at this long eared bat foetus. (figure 3) It looks uncannily human to me with those little feet!

There is so much more I could say about the delightful bat such as the fact that unlike other mammals they have developed mechanisms that make it possible for sperm cells to remain fertile over many months or that they can live ten times as long as a mouse, or even that bats separated from other mammals at the end of the cretaceous period some 70 million years ago. Interested? – Good, then off you go on your magical journey into the world of bats.

Bibliography

Bats of Britain, Europe & Northwest Africa by Dietz, Helverson and Nill
Bat Conservation Trust UK
http://www.crystalwind.ca/

Born in Birmingham, England Carl Portman has always followed the maxim 'interest is where you find it' and this certainly applies to natural history. He has bred endangered species of tarantula spiders, written two books on natural history travel, and lectures around middle England on animals and rainforests. Oddly he has a diploma in sexing juvenile theraphosid spiders, is an English Chess Federation County Chess Master, supports Aston Villa and has a strange addiction to Turkish Delight (covered in chocolate). Having worked for the Ministry of Defence for 30 years he now spends his time doing lecturing, chess coaching, some photography and management consulting.

He has spent time studying animals in the rainforests of Australia, Ecuador and Costa Rica searching for new and ever curious insects and arachnids and has a desire to find a new species somewhere in the world. His motto is 'Don't complain about the dark, light a few candles'.

He is married to Susan and lives in Oxfordshire. Their two Border collies, Darwin and Dickens keep them fit and ensure that there is never a dull moment in the household.

watcher of the skies

CORINNA DOWNES

Good news and bad news over these last few months.

Breeding Avocets

The good news includes the reports that avocets hatched young in Somerset for only the second time in 170 years; a brood of four chicks.

The chicks hatched on a specially-designed island at WWT Steart Marshes on one of several gravel topped islands that were designed by WWT to encourage ground nesting birds and funded by Taunton-based environmental trust, Viridor Credits Environmental Company, through the Landfill Communities Fund.

WWT reserve manager Alys Laver said: "*Steart Marshes was only finished last September, so for avocets to choose to breed here is quite a blessing and we're absolutely delighted! The local volunteer wardens have kept an eye on the family as mum and dad helped their chicks venture out for their first swim. Their island home is a little way away from the footpath along the bank of the River Parrett, but anyone should be able to see them with a pair of binoculars and a little patience.*"

WWT Steart Marshes is the result of the UK's biggest ever coastal realignment scheme, which helps manage flood risk for 100,000 homes and businesses in the Severn Estuary. The sea was allowed onto several hundred hectares of the Steart peninsula for the first time last September. Since then it is rapidly turning into saltmarsh, a habitat that's important for farming, fishing and wildlife but is under severe threat from rising sea levels.

Source: http://www.wwt.org.uk/news/all-news/2015/07/wwt-news/somerset-locals-celebrate-rare-chicks/
Photo: http://www.bto.org/

Bittern back to 19th Century levels

Bittern were extinct in the UK by the end of the 19th century and was absent as a breeding bird between the 1870s and 1911, when the first breeding male was recorded. It returned to peak numbers in the 1950s with around 80 breeding males, but then the decline began again, attributable to habitat loss. By 1997 there were only 11 breeding males recorded in England.

Concern over a second UK extinction led to a concerted conservation programme which is driving the current recovery.

Scientists count bitterns by listening for the male's foghorn-like booming song, and this year over 150 males have been recorded in England and Wales. Simon Wotton, an RSPB

conservation scientist, said: "In the late 1990s, the bittern was heading towards a second extinction in the UK, largely because its preferred habitat – wet reedbed – was drying out and required intensive management, restoration and habitat recreation. Thanks to efforts to improve the habitat, combined with significant funding from two projects under the European Union Life Programme, the bittern was saved, and we're delighted that its success keeps going from strength to strength."

Martin Harper, the RSPB's conservation director, added: "The bittern is a species which proves that conservation can be successful, especially when you can identify the reason behind its decline and bring in measures and funding to aid its recovery."

Over the last 25 years there have been several significant habitat-restoration projects, some of which are now RSPB nature reserves, including Ham Wall in Somerset, Lakenheath in Suffolk, and Ouse Fen in Cambridgeshire. According to this year's figures, the top UK county for bitterns is Somerset, with over 40 booming males.

East Anglia with over 80 booming male bitterns remains the bittern's regional stronghold in the UK, particularly in traditional sites on the Suffolk Coast, and in the Norfolk Broads but also increasingly in the Fens, particularly at newly created habitat. Over half (over 59 per cent) of the booming males are on sites protected under international law, namely the European Union's Birds and Habitat's Directives. These sites, referred to as Special Protection Areas or Special Areas of Conservation, are collectively known as Natura 2000 sites. Martin Harper said: "These sites have been vital to the conservation of the bittern and other key species in the UK. However, the European Union is consulting on the future of the Birds and Habitats Directives. And we fear this may lead to a weakening of the directives, with potentially disastrous consequences for many threatened species."

Source: http://www.rarebirdalert.co.uk/ ; http://www.wildlifeextra.com/go/news/bitterns-uk.html#cr

Photo: http://www.wildlifeextra.com/go/news/bitterns-uk.html#cr

Parrot Feared Extinct Is Found

The night parrot, believed extinct since the last confirmed sighting in 1912, has been found to be alive in arid desert in Queensland. The discovery of two dead birds between 1990 and 2006 only added to the belief that the parrot was no more.

But in 2013 wildlife photographer and naturalist John Young captured a few seconds of video footage of a live bird, which has now been backed up by the live capture and tagging of the elusive bird in a move that is seen as the "Holy Grail" for ornithologists.

It is so rare that ornithologists have kept secret the exact details of its location on the remote stretch of 56,000-hectare land where it was spotted and they are trying to raise funds to buy the land to keep their bird safe.

Night parrot expert Dr Steve Murphy, who helped to confirm the discovery, is thrilled. "I've been fascinated with night parrots ever since I was a small kid," he told Bush Heritage Australia. It's their story that grabbed me, and what it represented about what's happened to Australia since the arrival of Europeans. We've lost more native animals than anywhere else on Earth, and for a lot of years we thought we'd lost this one as well."

The editor of Birdlife Magazine Sean Dooley

summed it up as: "*The bird watching equivalent of finding Elvis flipping burgers in an outback roadhouse*". South Australian Museum collection manager Dr Philippa Horton called the find: "*One of the holy grails, one of the world's rarest species probably*".

Source: http://news.sky.com/story/1533820/ parrot-feared-extinct-for-100-years-is-found ; http://mobile.abc.net.au/news/2015-08-10/night-parrot-nature-reserve-created-queensland-endangered-bird/6680392?pfm=sm&section=nt
Photo: http://news.sky.com/story/1533820/parrot-feared-extinct-for-100-years-is-found

New Bird in China

A Michigan State University professor was part of an international team of scientists that has discovered a new bird in China; the Sichuan bush warbler, resides in five mountainous provinces in central China. Its distinctive song eventually gave it away, said Pamela Rasmussen, MSU integrative biologist, assistant curator at the MSU Museum and co-author on the paper. "*The Sichuan bush warbler is exceedingly secretive and difficult to spot as its preferred habitat is dense brush and tea plantations,*" said Rasmussen. "*However, it distinguishes itself*

thanks to its distinctive song that consists of a low-pitched drawn-out buzz, followed by a shorter click, repeated in series."

While the bird may be elusive, it is common in central China and doesn't appear to be under any imminent threat, she added. The bird's signature song can be found on MSU's Avian Vocalizations Center website. Thousands of bird songs are housed in this extensive website, including the new bird's closest cousin: the russet bush warbler.

Both warblers can be found on some of the same mountains. However, where they reside together, the Sichuan bush warbler prefers to live at lower elevations. When not competing with its cousin, the Sichuan bush warbler breeds up to 7500 feet.

Along with sharing the same mountain habitat, the two warblers also are close neighbours in terms of genetics. Analyses of mitochondrial DNA show that the warbler species are closely related and are estimated to have had a common ancestor around 850,000 years ago.

The bird's Latin name, *Locustella cheng*i, honours the late Cheng Tso-hsin, China's greatest ornithologist. Cheng, founder of the Peking Natural History Museum and author of 140 scientific papers and 30 books, was known internationally for his dedication to ornithology. *"We wanted to honor Professor Cheng Tso-hsin for his unparalleled contributions to Chinese ornithology,"* Rasmussen said. *"Many species are named for European explorers and monarchs but few bear the names of Asian scientists."*

Source: http://www.sciencedaily.com releases/2015/05/150501095953.htm
Photo: Laojun Shan

Now some of the bad news. Not only do the birds of prey killings go on but we have a new contender for the most-hated bird it seems.

Seagulls are on the most-wanted listed these days, especially ever since the report of the alleged killing of a dog by one. Retribution includes a new 'sport' whereby people deliberately catch seagulls using fishing lines, or seagull fishing as it has been dubbed.

RSPCA inspector Paul Kempson said: *"Deliberate cruelty is not only completely callous and unacceptable but it is also against the law. We have launched an investigation into the incident and would urge anyone with any further information to please contact our inspectorate appeal line on 0300 123 8018.*

"If you see happen to see anyone being cruel to a gull, please contact our national cruelty line immediately on 0300 1234 999."

Source: http://www.northdevonjournal.co.uk/ RSPCA-warns-latest-blood-sport-seagull-fishing/ story-27661319-detail/story.html#ixzz3kb MHbR4A

and then we have the 70-year-old woman being handed an ASBO for feeding seagulls in Devon:

A council in Devon has taken the unusual step of slapping an ASBO on a pensioner for feeding bread to seagulls. The 70-year-old has been banned from feeding seagulls and all other birds in her home town of Sidmouth, but the local council banned her after complaints from some residents.

She has even been threatened with eviction from her council home if she continues feeding the gulls, pigeons and doves. Rose believes that concerns raised from big hoteliers in Sidmouth town about the "flying rats" are behind her "victimisation" from the council. She said: "*I have been feeding them for three years - and they initially found me. The ban is to stop feeding any birds in the whole of Sidmouth - I am not even allowed to feed them in my own garden. This all stemmed from my neighbours complaining - they hate anything in their gardens - even a hedgehog that is there that I am not allowed to feed. But I believe the real problem in Sidmouth is the big hoteliers. They don't want pigeons and seagulls and this is why the council has gone after me.*

"*I was given 21 days to appeal my mini ASBO, which is what I have done. But they are digging up my whole history. How can I upset the local residents just by feeding pigeons in the cemetery? I initially thought the ban was for everyone feeding them - I could not believe it was just me. I have just turned 70, but they have even threatened me with eviction and say I will be in breach of my tenancy in my council cottage if I continue feeding the birds or the hedgehog. There is no-where else for me to go. People are just so intolerant these days.*"

Rose, originally from London, said she has always been a bird-lover but only started feeding them in Sidmouth three years ago when she found a few injured homing pigeons in the town's Market Square, and she fed the ones that needed help. She said: "*I am fighting to get it overturned and I have been granted legal aid so we will have to see what happens.*"

Three years ago, East Devon District Council also gave Rose an £80 littering fine for feeding birds in the town. She was given the fixed penalty notice while feeding peanuts to pigeons.

Giles Salter, representing East Devon District Council, told the court he believed that Rose's appeal was the first against such a notice in the country. The matter will come before the court again in October after both parties have had chance to put their cases together.

Source: http://www.northdevonjournal.co.uk/ASBO-handed-woman-70-feeding-seagulls/story-27665456-detail/story.html#ixzz3kbMpf4kJ

Then there was the case of mistaken identity over two birds that look very similar: Pukeko culls halted after endangered takahe shot

Pukeko

Takahe

The New Zealand Department of Conservation has put an immediate halt to pukeko culling operations near threatened takahe populations after four takahe were shot in a cull on Motutapu Island in the Hauraki Gulf. DOC confirmed this morning an examination of the dead birds on the island sanctuary early this week showed they were killed by shotgun pellets.

The island's cull was undertaken by "experienced members" of the local deerstalkers association, DOC's northern conservation services director Andrew Baucke said. DOC was in talks with the association, which is said to be "co-operating fully" with inquiries.

Mr Baucke said takahe and pukeko had similar colouring and could be mistaken for each other. The hunters were carefully briefed on how to tell the difference between them, including instructions to only shoot birds on the wing, he said. "*Guidelines introduced after an incident on Mana Island seven years ago when another takahe was mistakenly shot during a pukeko cull were also used during last week's cull.*"

Despite long-running conservation efforts, there are only 300 takahe left and the species is classified as "critically endangered". The takahe on Motutapu had been translocated from the Fiordland National Park, where the only wild population of the birds is based.

Pukekos are as common and ducks and geese and because they are a highly aggressive species they are considered a threat to rare native bird species. Takahe were thought to be extinct in the early 20th century but were rediscovered in 1948 in the South Island. Two-thirds of the population are now based in "safe sites" including Motutapu, while around 100 live in the wild within Fiordland National Park.

DOC, in partnership with Mitre 10, has a goal of establishing 125 breeding pairs by 2020. The public-private conservation programme has been running for 11 years at a cost of $292,000 to DOC and $150,000 to Mitre 10.

Takahe
Population: 275-300
Habitat: Fiordland National Park (wild). Maud, Mana, Kapiti, Tiritiri Matangi and Motutapu islands (sanctuaries).
Appearance: Dark blue head and neck, turqouise and green back and wings, red beak and legs.
Adult size: 50cm long, up to 3kg

Pukeko
Population: Abundant
Habitat: Marshy roadsides and low-lying open country throughout New Zealand.
Appearance: Deep blue colour, with black head, red bill and legs.
Adult size: 50cm long, 1 to 1.5kg
- Department of Conservation

Source: http://www.nzherald.co.nz/nz/news/article.cfm?c_id=1&objectid=11500322
Photos: https://en.wikipedia.org

Other news:
Bee-eaters breeding - Cumbria
The announcement on 31st July of bee-eaters breeding, with up to six birds at two nests near Brampton in Cumbria, was an exciting declaration. This is the second year in a row in which bee-eaters have bred in this country; last summer's birding news was brightened considerably by the goings-on on the Isle of Wight, where two nests successfully fledged eight baby bee-eaters from their chosen site on the Wydcombe Estate.

The most recent known breeding attempt came in the summer of 2005 in Herefordshire, but was unsuccessful due probably to fox predation, whilst three years earlier, Britain's second successful breeding record came from County Durham resulting in two from a brood of five fledging successfully.

Prior to that, two out of three pairs raised seven young in East Sussex in the summer of 1955 while another failed attempt was recorded in Scotland in 1920, when a pair nested along the banks of the River Esk near Mussleburgh, but the female was captured by a local gardener who kept her in a greenhouse where she died a couple of days later, after

laying one egg.

Source: http://www.rarebirdalert.co.uk/
http://www.birdguides.com/
Photo: https://en.wikipedia.org

On 18th August the first fan-tailed warbler for Sussex, and the first in Britain since 2010 was seen briefly at Climping. There are just 10 records to date: Ireland - both in April at Cape Clear - and a further eight from Britain, in Kent, Dorset and Norfolk. This Old World warbler's breeding range includes southern Europe, Africa (outside the deserts and rainforest), and southern Asia down to

northern Australia.

Source: http://www.rarebirdalert.co.uk/
http://www.birdguides.com/
Photo: https://en.wikipedia.org

At the end of July/beginning of August, the first Caspian tern for Essex in 34 years was recorded at the scrape at Holland Haven CP. But not only did it become the first for over three decades in the county, it was also only the 7th Essex record in all - although two old records, both from 1961 (at the reservoirs of Walthmstow and King George V) should arguably moved to "London" ~ which then takes the total for Essex down to just four in all prior to this week's bird. Singles have been seen at Abberton in August 1959 and August 1961 and others appeared at Colne Point and Fingringhoe Wick in June-July 1975 with the most recent record until now coming from Heybridge GPs on June 20th-22nd 1981. It is the world's largest tern and its breeding habitat is in North America, and locally in Europe (mainly around the Baltic and Black Seas) Asia, Africa, and Australasia. North American birds migrate to southern coasts, the West Indies and to northernmost South America. European and Asian birds spend the

non-breeding season in the Old World tropics. African and Australasian birds are resident or disperse over short distances.

Source: http://www.rarebirdalert.co.uk/
            http://www.birdguides.com/
Photo: https://en.wikipedia.org

The first Hudsonian godwit for Ireland was discovered on 22nd July at Ballyconneely in County Galway. This mega vagrant breeds in the far north near the tree line in

northwestern Canada and Alaska, as well as

on the shores of Hudson Bay. They migrate to South America, and it is not unknown for them to be a vagrant to Australia and South Africa.

Source: http://www.rarebirdalert.co.uk/
        http://www.birdguides.com/
Photo: https://en.wikipedia.org

An unusual bird in a Shetland garden proved to be the archipelago's fourth eyebrowed thrush on Whalsay on 4th July, and the first away

from Fair Isle and Foula. Following last year's bird on Orkney's North Ronaldsay, this latest bird takes the British total to 22 individuals. After the bout on Scilly in the late 1980s and early 1990s this species has reverted to being a genuinely rare bird.

Source: http://www.rarebirdalert.co.uk/
        http://www.birdguides.com/
Photo: https://en.wikipedia.org

On 10th June, a Cretzschmar's bunting on Bardsey, Gwynedd, was the sixth British record, but the first away from the Northern Isles. They have been seen on the same date for the last 48 years: in 1967, 1979 and 2015. It breeds in Greece, Turkey, Cyprus and the coastal countries along the eastern edge of the Mediterranean. It is migratory, wintering in the Sudan. It is a very rare wanderer to western Europe. The name commemorates the German physician and scientist Philipp Jakob Cretzschmar who founded the Senckenberg

Natural History Museum.

Source: http://www.rarebirdalert.co.uk/
http://www.birdguides.com/
Photo: https://en.wikipedia.org

On the same day, a cedar waxwing visited on Tiree in Argyll, which was the sixth record in the British Isles - although the second for that island. The cedar waxwing is a common North American bird that winters south to Central and northern South America. Most of the population migrates farther south into the United States and beyond, sometimes reaching as far as northern South America.

Source: http://www.rarebirdalert.co.uk/
http://www.birdguides.com/
Photo: https://en.wikipedia.org

On the 9th June, in Ireland, a slate-coloured junco arrived on Dursey Island, County Cork, which is the fourth national record. There was also a male on the southern end of Mainland - the fifth for Shetland and the first since May 2003. The first for the archipelago was found in 1966 on Foula and was followed, just over a year later, by another male on the same island. Shetland's third made itself known on Out Skerries in 1969. This week's becomes the 45th record for Britain and Ireland and the first

one since the wintering male in the New Forest, at the Hawkshill Inclosure between December 2011 to February 2012. They breed in North America from Alaska to Newfoundland and south to the Appalachian Mountains, wintering through most of the United States.

Source: http://www.rarebirdalert.co.uk/
http://www.birdguides.com/
Photo: https://en.wikipedia.org

A veery on North Ronaldsay on the 30th May was the second record for Orkney and the 11th for Britain. The only previous spring record of

this species concerned a bird on Lundy in Devon on 14th May 1997. Coincidentally the archipelago's only other record, from 30 September 2002, came not only from the same island but also the same garden! The veery breeds across eastern North America and winters over a large area of northern South America.

Source: http://www.rarebirdalert.co.uk/
http://www.birdguides.com/
Photo: https://en.wikipedia.org

A grey-cheeked thrush at Termoncarragh, County Mayo was a much unexpected discovery on 25th May. This is the first spring record for the British Isles, and is only the second of the nine Irish records to occur outside County Cork. Their breeding habitat is the northern spruce forests across northern Canada and Alaska. Their breeding range extends into Siberia. These birds will migrate to northern South America. This species is a rare vagrant to Europe.

Source: http://www.rarebirdalert.co.uk/
http://www.birdguides.com/
Photo: https://en.wikipedia.org

A pair of hooded mergansers were discovered on Tory Island, County Donegal during mid-May. If confirmed as wild birds, they will be the fourth and fifth individuals for the Republic of Ireland, following one off Ballylongford, Kerry in 1881

and two in Cork Harbour in 1878. Another was in County Armagh, Northern Ireland in 1957. This sexually dimorphic small duck is the second smallest species of merganser, with only the smew of Europe and Asia being smaller, and it also is the only merganser whose native habitat is restricted to North America. Although the hooded merganser is a common species in captivity in Europe and most specimens recorded in the wild are regarded as escapes, a small number of birds have been regarded as genuinely wild vagrants.

Source: http://www.rarebirdalert.co.uk/
http://www.birdguides.com/
Photo: https://en.wikipedia.org

The first Moltoni's subalpine warbler that has ever been reported for mainland Britain was discovered on Blakeney Point, Norfolk on 11th May. The first accepted Moltoni's subalpine warbler was shot on St Kilda in 1894, with two more recent records in Shetland in 2009. Further positive records include two sound-recorded males - both on Shetland in the spring of 2009, at Scatness in May and on Unst, at Skaw during June, which were numbers 2 and 3 for Britain. This bird breeds in Italy, Sardinia,

southern France, southern Spain, and the Balearic Islands.

Source: http://www.rarebirdalert.co.uk/
http://www.birdguides.com/
Photo: http://uk400clubrarebirdalert.blogspot.co.uk/

Britain's second citril finch was discovered in Norfolk on 10th May in the dunes at Burnham Overy. This male is only the second British record in nearly seven years after the discovery of one on Fair Isle in early June 2008. For almost 90 years the species' place on the British List came courtesy of a "citril finch" that was taken alive on the North Denes, in Yarmouth in late January 1904. That bird was soon admitted to the British List and was finally photographed in 1989 by Yarmouth's legendary birder Peter Allard who was researching a book on the local area's avifauna. Still in the original case, the "citril finch" rang alarm bells. The images he took of the skin eventually found their way to Lee Evans, who immediately suspected a canary of some sort was involved and then Dr. Alan Knox (then representing the BOURC) was made aware of the situation and, after visiting the Booth Museum, he removed the specimen to the British Museum in Tring, where finally Yarmouth's "Citril Finch" identity was confirmed - it was actually a male Cape canary. Hence, the Citril finch was officially removed from both the Norfolk and British Lists in 1993 – but now, 22 years on it can be re-added. This bird is a resident breeder in the mountains of southwestern Europe from Spain to the Alps. Its northernmost breeding area is found in the Black Forest of southwestern Germany.

Source: http://www.rarebirdalert.co.uk/
http://www.birdguides.com/
Photo: https://en.wikipedia.org

For those of you not aware, as well as this column in *Animals & Men*, Corinna writes a daily Fortean bird blog which can be found as part of the CFZ Blog Network, but also as a stand alone site at:

http://cfzwatcheroftheskies.blogspot.com/

HUNTING FOR LOST ANIMALS A LONG WAY AWAY

In some ways Australia is the perfect place to go on an expedition. The whole thing is so bloody far away that just getting there makes you feel like you have already accomplished something. You arrive with a sore bottom, slightly delirious from lack of sleep, with breath like a hungover dragon, but convinced that nothing is impossible. You found Australia, so of course you can find the thylacine!

That was more or less my state of mind, when I arrived in Sydney Airport in the beginning of February this year to take part in the second CFZ expedition to Tasmania to hunt for evidence of the continued existence

L-R: Lars, Mike, Chris, Rebecca, Tony

A report from the second CFZ Thylacine expedition to Tasmania – February 2015
by Lars Thomas

of the thylacine, or Tasmanian tiger if you prefer. This weird marsupial has clocked up some 4000 sightings despite having been deemed extinct since the 1930s. Actually it has only been officially extinct since 1986, because that was the 50th year since the last officially excepted sighting (in 1936 of an animal in Beaumaris Zoo in Hobart), and that as it were, is the internationally agreed way of deeming an animal extinct.

But the point is of course – is officialdom and science wrong? It wouldn't be the first time, and with so many sightings logged, it is extremely tempting to think that in this case science is wrong big time!

Well, the only way to check up on science is to take a look for yourself. So in 2013 CFZ sent off their first expedition to Tasmania. They turned right, when they arrived in Tasmania and had a look on the Northwest part of the island. This year, we went to the northeast to see what was going on there. According to veteran researcher Tony Healy, who had been sent on ahead as a scouting party, quite a lot – plenty of local sightings just within the last decade.

Getting started
As it turned out – this was to be an expedition besieged by a number of difficulties – but more of that later. It started well. I was the first to arrive at CFZ-Australia HQ, i.e. the house of Rebecca Lang and Mike Williams in the Blue Mountains west of Sydney. I was a couple of days early, which was great, because it gave me a chance to get my bearings before going south to Tasmania. So while Rebecca and Mike was busy rounding up vehicles and getting our stuff (and act) together, I was left to my own devises, and consequently pottered around the local area watching cockatoos and parrots, collecting the odd big cat sighting 200 meters up the road from the house, and generally having a nice and relaxed time, when I was not being stared at in a reproachful way by the strangely silent CFZ-Australia basenji Jett, who had a disconcerting habit of sneaking up on me, and just standing there staring, making me feel like an uncouth intruder of the worst kind.

But all went well, and off we went, driving across New South Wales and Victoria to Melbourne to sail across the Tasman Sea to Devonport. Along the way I sampled a few of the local tourist-attractions, such as being photographed leaning on the biggest set of ram's testicles I have ever seen (don't ask – but in my defence I will say they were made of concrete). After an overnight stay somewhere in the middle of it all, we arrived in Melbourne at the ferry (having negotiated a monumental traffic jam), and sailed off into the wild, blue, and severely choppy yonder.

Despite having spent most of my life studying animals of various kinds, I am not – repeat not – a morning person, so when an obnoxiously energetic voice on the ship's loudspeaker system woke me up the next morning at 6 o'clock to inform me that the boat had arrived at Devonport Harbour, and that I should basically get a move on, I felt a sudden urge to do bloodshed in the time-honoured custom of my forebears. But, being fairly civilized, or at least something along those lines, I got up, and grumbled my way to our cars in the ship's hold. The weather was grey and rather dismal, but luckily all we had to do for that day was to meet up with Tony Healy and find the cottage that was to be our Command Central for the expedition.

Our first mission of the day was

accomplished in the little town of Scottsdale, where breakfast was taken, and I met up with the legendary Tony Healy, affectionately known as the Healy Monster, driving in his equally legendary Healymobile, a converted bedroom, laboratory, file and general utility van. Our second mission took a little bit longer. We had to drive towards the northeast, passing by the rather disturbingly named Mount Horror, until we got to South Mount Cameron, which is A: a mountain, B: the general area around said mountain, and C: a small village. Our cottage was located at the foot of A, in the middle of B, and just outside C.

Here we sat up shop, started making plans, and waited for the rest of the expedition – Dr. Chris Clark and Richard Freeman. The latter was unfortunately never able to make it, as he went and got himself ill almost on the day of departure. Chris Clark made it – flying in from France – but was only able to stay for part of the expedition.

The Weld Valley and the Abominable Journalist
Plans were also a foot for an Australian journalist and a photographer to spend a couple of days with us, with the intention of writing an article for the *Sydney Morning Herald* about the expedition. That turned a bit sour, as everybody was led to believe the article would be rather serious and scientific, and it turned out to be making rather severe fun of the whole setup. Luckily the article didn't come out until a couple of weeks after everybody had returned home, as it would have had a rather dampening effect on the expedition morales.

On the other hand, for the benefit of same journalist, who shall remain nameless, a trip was arranged to take Mike Williams and yours truly, as well as the journo and a photographer for a one night stay at the edge of the Weld Valley, the area where Col Bailey, one of the most legendary thylacine-hunters had his own meeting with a thylacine, something which has been described in detail elsewhere, so no need to do it again here, although the valley is quite a magnificent wilderness. It is incredible that Col Bailey, or anybody else for that matter, has been able to penetrate it. Not only because it is so dense, but also because of a rather large population of tiger snakes, and *those* you do not want to get up close and personal with, especially not if you are on your own, somewhere way out in the wilderness.

The Weld could easily be home to a number of thylacines, but due to time and equipment constraints, we could only scratch the surface as it were. An automatic camera that went up for the night did not catch anything, but I managed to find several samples of hair caught on branches near what were very clearly animal track ways through the undergrowth. Unfortunately none of the hairs turned out to be anything out of the ordinary.

The wonderful witness
We had relied rather heavily on Tony Healy to line up eyewitnesses for us to talk to. And he had really pulled it off with retired police officer Richard Compton who could boast no less than three separate thylacine sightings. At first he was rather suspicious of the whole thing, feeling that the animal would probably be better off living off the grid, as it were. Something I can't entirely blame him for thinking, as the track record of the Tasmanian authorities when it comes to protecting their various wilderness areas is less then exemplary. But in the end he relented – probably because of Tony Healy's extremely sensitive and diplomatic

behaviour, and told the story of his sightings – one of which actually included him taking an entire roll of film of an animal standing in the middle of the road staring at him – without removing the lens cap!! As such, they sightings didn't really contribute anything new to our knowledge of the thylacine, but the man himself was something else. Tall, grizzled and tough as nails, he looked exactly like an older version of Crocodile Dundee. His stories were quite straightforward, but following our excursion into thylacine territory, he started regaling us with stories of his exploits as a cop. And they were, shall we say, rather colourful. Somebody should make a movie about that man!

Interesting as Richard Compton was, he was in a sense just a minor part of the expedition. What we really wanted to do, was to search the Northeast part of the island, and that was what we actually spent most of our time doing.

Mt. Cameron
The area around Mt. Cameron where we were based should, in theory, be perfect for any medium-sized predator. There are large areas of extremely dense scrub and forest around the mountain, and large open areas with coarse grass and scattered bushes – all with a healthy population of tiger snakes, and a very large population of small and medium-sized marsupials – and a few big ones as well. Sheep and cattle were also to be found, but not in as great numbers as one would expect. The human population is also fairly thin on the ground around here, so the area is fairly peaceful as well.

So – plenty of hiding places, good hunting grounds and probably not much pressure from humans. And a lot of sightings – especially by people driving through the area at night, and suddenly seeing a thylacine standing in their headlights, giving them the onceover before disappearing into the brush.

We spent a fair bit of time scouting the place, trying to find suitable areas for automatic game cameras, as well as experimenting with various forms of baits and lures to get animals to move in and have their portraits taken. This also meant scuttering round the roads at night with big black bin liners gathering roadkills. Luckily we weren't spotted by anyone – it would have been necessary to do quite a bit of fast talking if so.

The cameras did not produce any usable results thylacinewise, although we did get shots of several different species of marsupials, including Tasmanian devils, which was rather good, as our study-area is located in the middle of one of the worst hit areas by the DFTD (Devil Facial Tumour Disease), so one could have feared that the devils had been totally eradicated, but luckily no. The main reason we did not get much in the way of results was probably partly because we only had a fairly limited amount of time and a limited amount of cameras, meaning we could only cover a very small area. We did get some interesting shots of quolls and a few Tasmanian devils that luckily seemed to be completely healthy – one of them made no secret of the fact that it had seen the camera all right! But despite baiting some of the cameras with juicy freshly collected roadkills – not even the hint of a trace of a thylacine.

Another technique we employed was nightly spotlighting runs on the various roads of the area. Since the vast majority of the thylacine sightings are of animals suddenly appearing in people's headlights after dark, we also employed dashcams that recorded everything

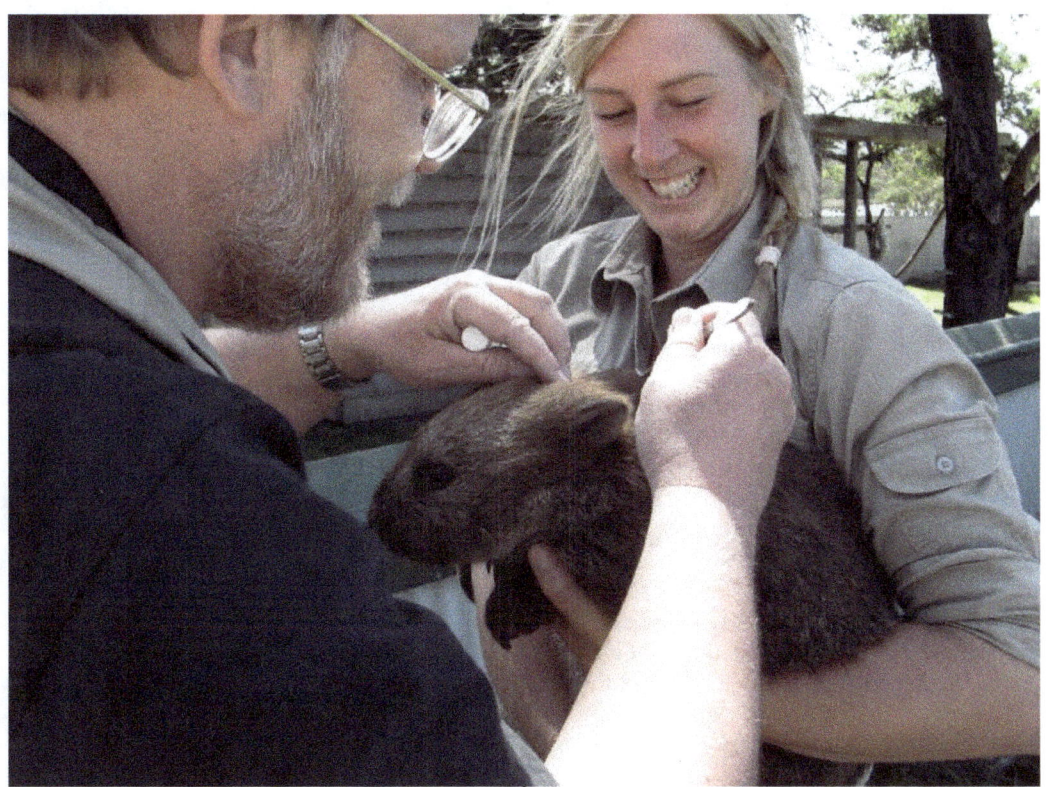

in front of the car. An added bonus of the driving was the chance to assess the general population of marsupials in the area, which as mentioned above is fairly large, and thus get a rough idea of whether it would actually sustain a population (although small) of larger marsupial predators. So, every night I would count the various animals we did see (because we didn't see any thylacines). The list of species was actually fairly impressive:

Short-beaked Echidna
Tiger Quoll
Eastern Quoll
Tasmanian Devil
Common Wombat
Common Brushtail Possum
Southern Bettong
Long-nosed Potoroo
Rednecked Wallaby
Eastern Grey Kangaroo (or Forester Kangaroo, as it is known in Tasmania)
Tasmanian Pademelon

Hairs, hairs, hairs and more hairs – and some scats too!
In between all the spotlighting and other suspicious looking nightly activities, I also, thanks to Rebecca Lang's oratorical gifts, was able to considerably extend my reference library of hair samples for future studies – after all, I only need one tuft to prove the Thylacine is still around – but I also need to know what the hairs of all the other animals looks like. Rebecca Lang somehow persuaded the people at the

Bicheno Wildlife Park to let me take hair samples of their various animals, something they took in their stride. Although they did get a slightly haunted look when she wanted scat samples as well. But samples we got, and spent an interesting, and slightly smelly afternoon, photographing the scats. After all, it's nice to be able to eliminate the known species, should someone in the future find something interesting .

A sideorder of bugs, spiders and other creepy crawlies
In an effort to build a positive relationship with some of the local museums and scientists, I offered to collect insects and other invertebrates for the museum in Hobart after reading that the pitifully few zoologists working in Tasmania reckoned that they only knew about 1/3 of the insects species in the state, and that field guides and identification keys was something they could only dream about. So, after applying for, and getting an official collectors permit, I kept an eye out for whatever critters I could lay my hands on.

This will not be a detailed description of this mainly entomological part of the expedition, but just to give everybody a taste of some straightforward natural history, what follows is a small selection of what I actually caught.

This blue cuckoo bee which I found dead in one of the windows of the cottage we were staying in, has so far eluded all attempts of identification, and may be a new species.

Next a known species - but a very nice looking ones. Cleobora is an Australian ladybird.

This tiny jumping spider has so far refused to be identified, and might be another new species.

This nice looking bug is also proving elusive as to identity, and may possibly be a new species as well.

As shown above, some of the animals haven't been identified yet, and there is a fair chance that a few of them are actually new to science. Who knows, someday a small bug may be crawling around the Tasmanian bush sporting a CFZ inspired scientific name!

The future
We did, as it were, not add much to the knowledge of the Thylacine, but it became quite clear, that further expeditions have to be done on a much larger scale, if anything is to come out of all this. For a small group to actually meet up with a Thylacine, they would need to be beyond lucky.

I am tempted to write several puns along the lines of needles in haystacks, thylacines in bush, and so forth, but I shall refrain from doing so. It would probably take a large number of game cameras in a large area for a long time, to come up with something interesting. Or an equally large number of vehicles supplied with dashcams, driving all over the island, or at the very least doing nightly patrols in the same area for several weeks on end. There are now so many sightings recorded that collecting more doesn't really add to our knowledge. We need to get footage or physical remains, and that takes a lot of manpower, and a lot of resources.

And finally, we need to seriously consider the effect of the extreme exposure the Thylacine gets in modern culture in Tasmania.

The animal is absolutely everywhere – in commercials, in logos, in newspapers, on toys, on lunchboxes, on bus shelters, on letterheads, on caps, on T-shirts, on ice-cream, on just about everything. The image of the thylacine is so engrained in everyday life that it has probably generated a substantial number of the recorded sightings.

A LITTLE KNOWN BRITISH PSEUDOCRYPTID: THE MONGOOSE

There is a little known British pseudocryptid, the mongoose, encounters with which reached its heyday between about 1905 and 1910, though there are cases on either side of these dates.

They ranged from Trowbridge in Wiltshire to the Isle of Bute on the west coast of England. There was a proposal to introduce them as late as 1929 to control vermin, but prior to this they in turn were hunted on occasions by ferrets. What happened to them after 1929 I do not know. It was a *proposal* not a definite accomplishment.

There is a document at the National Archives in London dated that year about this proposal. [1] The trail seems to go cold between about 1910 and 1930 concerning what happened to Britain's mongooses and afterwards.

No examples of British mongoose at all can be found in the General Index to *Fortean Times* nos. 1-66. But if the earlier introduction at the beginning of the 20th century failed, why did the powers that be try it again and what happened to the first "wave" of imported British mongooses? If the geographical range of their introduction was so widespread, what happened to them all? Presumably they died out so quickly because they were only ever introduced singly and never in pairs of male and female and therefore there were no offspring. Or perhaps they were soon considered vermin themselves?

Mongooses were introduced to Trinidad and Jamaica in 1800s to control the rat population in the sugar cane plantations and would have been imported to Britain from places like India, and parts of Africa including Egypt and South Africa. Maybe even Hong Kong where there are two species of mongoose. They have also been introduced to Puerto Rico Hawaii and Fiji.

It has been suggested by some Fortean zoologists that some British big cats are descendants of ones brought into Britain during World War II, this being about forty years before the "modern" phase of ABC reports really "taking off" in the 1980s. But on less slender evidence, the mongoose in the UK has been neglected or ignored. I admit I am guilty of this too, despite finding a case of one in Northamptonshire (see below) when I was in hospital there briefly in August 1997 and went to the town library. As early as 1885 the *Manchester Courier* and *Lancashire General Advertiser* for 6 August 1885 p. 3, reported:

> "THE INDIAN MONGOOSE.
> The *Daily Telegraph* says:—Is the mongoose an efficient substitute for the ferret? Here is a question which would appear to be agitating the gamekeeper mind, and, yet, emphatically the replies to are all in the affirmative. Not only it good, it is better, " The rabbits," says one expert "are seen to bolt more freely from him than from the ferrets- perhaps from the peculiar noises he made when put into the burrow. He never hid up or ate or damaged his rabbits, and we could always tell if there was a rabbit in the

Richard Muirhead

LE VANSIRE.

burrow the noise he made." This is delightful passage, and, as illustrating the results of a purely technical, practical consideration of a question having very wide "bearings," is altogether precious. The rabbits, we are told, seemed to bolt from the mongoose "more freely" than from ferrets and the reason alleged is the " peculiar noise made by the intruder. Now, ferrets are bad enough but after all Bunny knows what it is when he sees one. He has seen the weasel, no doubt, hunting copse-edge, and stood up on his hind legs with its inquisitive ears to watch the "lean, hungering, stoat" as it passed down the path the meadow. So the ferret is more or less "a common object of the country" to the British rabbit. But of the mongoose he knows nothing, and its appearance in its burrow is surely sufficient to make him bolt " freely." The apparition of the animal with its long bushy tail, and chittering" to itself as it goes, is surely enough to terrify any warren."

The next report I have is from the Welsh newspaper the *Evening Express* for January 18th 1894:

"There was a very peculiar incident occurred in a Cardiff furnisher's shop the other day that has since provided considerable food for discussion. At Messrs. Trapnell and Crane's premises an animal was heard running about under a sort of platform, that at first was thought to be a rat, then a cat. but when a glimpse was got of it no one knew what it was". A couple of fox-terriers got in to t-y and dislodge or kill it, soon had enough of the fray, and retired wore ted.

Then Mr. Roberts, a local fancier, dropped in, and, on the situation being explained, he volunteered to send for a Bedlington`s bitch that he had always found would "tackle anything moving on four legs "- that is of course, of a decent size. Well, as fortune would have it, to cut a long story thick, the dog arrived and drew the "thing." There was a bit of a scramble, and then Bedlington`s dog, nipping the mongoose across the back, did the trick. Then the question arose,

What was it? No one had ever seen a thing like it before, and it was only on consulting an expert that it was found to be a mongoose. Now, a mongoose is a living terror," and known in India- its home as a snake killer, but how, in the name of all that`s great, it got to Cardiff no one known, There has been. no show in the district from which it could have escaped, and the only explanation that can possibly be given is that it came over with some foreign bedding. Still, we all ought to be thankful to that Bedlington for settling its hash."

The following day the same newspaper reported:

DOG VERSUS MONGOOSE.
A CORRECTION
TO THE EDITOR OF THE "EVENING EXPRESS."

Noticing in your last evening's issue, under the heading of Dog versus Mongoose," an incident that occurred recently at our establishment, we write to say that the explanation you give of the mongoose having probably been imported with some foreign bedding is entirely erroneous.

We neither import any bedding materials or the manufactured article from abroad. We cannot explain how the mongoose came upon our premises, but we may add that we have since learned that a person residing in the neighbourhood keeps one of these animals as a pet. We will thank you to insert this correction in your earliest edition, as the report is likely to injure our large bedding trade.

We are,
TRAPNELL and GANE.

Animals & Men — Issue 54

38 and 41,
Queen-street,
Cardiff,
Evening Express January 19th 1894.

More stories appeared over the next few years:

Hartlepool Mail 14th July 1897 p. 3
MONGOOSES FOR RATS.
The *Newcastle Journal* says Mr Gaher, a well known fruit merchant in Newcastle Market, has taken up arms against the vermin by introducing mongooses to destroy them.

A mongoose will kill rat in a twinkling, in daylight or in dark, crushing its skull with the first squeeze of its jaws as it crushes the skull of a snake. The very presence a mongoose is enough to frighten rats away. Mongooses have been employed to clear the Messrs Bainbridge's establishment of rats, and they have rendered a like service several other business places in the town within the past few years. They should be kept in a warm place, especially the cold season approaches, for frosty night is death the exposed mongoose.

Hull Daily Mail
20 December 1898 p. 4
THE OWNER OF THE MONGOOSE. The mongoose which Inspector Wardall and Sergeant Dickinson found on Anlaby-road. yesterday morning, was this morning claimed from the Parliament-street Police-station. It had escaped from a cage Mr Addy's, Upper Union-street.

Bristol Mercury
16 May 1900 p. 8
A raid on rats is contemplated. Many unkind things are said about them, and now we are told that they are an important factor In the spread of plague. The Medical Officer of Health (Mr Davies) brought this fact before

the notice of the Health Committee yesterday. He pointed out that precautions are taken as to the inspection of steamship passengers, as to isolation and disinfection in regard to ships, and that attention to general sanitation in s regard to houses must necessarily be continued, and he strongly urged the consideration of special measures to lessen the prevalence of rats in and about riverside warehouses where chance introduction of plague might occur. The experience of India, and later of Australia, he added,

In his *Zoology Jottings* blog of June 17th 2015, Malcolm Peaker (who happens to be an authority on the zoology of Hong Kong, another keen research interest of Jon Downes and myself) placed a story for all to see online titled `The Isle of Bute Mongoose of December 1900`. [(2)] Part of the text is as follows:

> "Bute, an island in the north of the Firth of Clyde, is the last place one would expect to see a mongoose. But while looking for something else I came across this report of a mongoose found on the Isle of Bute in 1900 and even more remarkably of their employment in bakeries in Scotland as mousers. MONGOOSE Herpestes ichneumon
>
> An old record of a possible Mongoose on the Island of Bute has recently been discovered and reported (Gibson, 2000). Searching the columns of any local newspaper in the hope of discovering some interesting natural history items is a wearisome and usually unrewarding task, but just occasionally one finds something unusual, and in the Buteman for Friday 5th December 1900 I was fortunate enough to discover the undernoted paragraph:
>
> "That well-known trapper, Mr. Robert Morrison, had the luck to trap on Saturday

last, at Plan Farm, south end of Bute, an animal certainly not native of this country. It is believed to be a mongoose, and was caught in a trap set in a rabbit hole".

So far I have been unable to trace any later reference to this occurrence, but the Mongoose is a fairly distinctive creature and there is no real reason to doubt the identification, even although no confirmatory details were given. At first, it might reasonably be assumed that this occurrence resulted from another introduction, as yet unrecorded, by the Bute family, but the Bute family Archives were very thoroughly searched by the late Marquess and myself when we were endeavouring to obtain as much information as possible about the Wallaby and Beaver introductions, and nowhere in the Archives did we find any indication at all that the Mongoose was ever introduced.

There is, however, a much more likely explanation. In my 1976 account of the land mammals of the Clyde area (Gibson, 1976a) I reported the old records of an adult Mongoose trapped at Blanefield, West Stirlingshire, on 1st June 1928, and five weeks later a a barely half-grown specimen trapped at Duntocher, Dunbartonshire. At that time, the origin of these animals was unknown, but eight years later, by the time of the publication of my separate account of the mammals of Dunbartonshire (Gibson, 1984), my colleague Mr. John Mitchell had discovered that "in the 1920s Mongooses were kept by at least one Dunbartonshire bakery, since their mousing ability was considered to be greatly superior to that of cats" (Mitchell, 1983). Later investigation showed that this practice was more widespread, in several parts of the country, than had previously been realised.

It seems most likely, therefore, that the 1900 Bute Mongoose was an escape from a local bakery or some other establishment, but it is clearly desirable to draw attention to this occurrence in the hope that some other records may come to light, and needless to say, I shall be most grateful to receive any additional information.

I do not know whether Dr Gibson did receive more information but is I think it more likely that the mongooses brought to Scotland would have been the Indian Grey (*Herpestes edwardsii*) rather than the Egyptian (*Herpestes ichneumon*) since the former were commonly imported up to the time quarantine for rabies was introduced for a wide variety of mammals in 1974."

Mongooses were also often brought back by sailors as pets and it is no coincidence that the major animal dealers were located in seaports like London, Liverpool and Glasgow. The Indian Grey (and, I read, the Egyptian) become extremely tame very quickly. I had one for several years that was completely trustworthy when being handled; its delight in life was being given an egg. Usually it would remove the shell from one end before licking out the contents. Less frequently it would throw the egg down onto the ground until the shell shattered and the contents could be licked up.

Indian Grey Mongooses introduced into the West Indies and Hawaii have had a devastating effect on the native wildlife. I wonder if the mongooses kept in bakeries ever bred. Mongooses had the reputation of needing to be kept warm in Britain, so it is perhaps unlikely that they would breed, or even survive for very long, when feral in the cold, damp winters of the West of Scotland."

Another possible source of a escaped mongoose on Bute could have been the Royal Aquarium in Rothesay. They advertised a menagerie in addition to aquatic exhibits.

But there is nothing to beat seeing mongooses in the wild, from seeing them walking in pairs along the footpaths in a hotel grounds in Goa to their pottering about the forest in Gir Forest."

The controversy continued with the following article:

Evening Express 12 Jan 1901

"MEANDERING MONGOOSE. Apropos a note in our "By the Way" column to-day the following par from the "Mail's" Wales column comes in quite pat as illustrating how hard the night staff sometimes find it to work up excitement. A score of ferocious mongoose (or is it mon-goses or mongese?) might invade this office in the day time and we should not turn a hair.

Excitement reigned yesterday afternoon in these offices. It appeared that the mongoose which is retained at the Conservative Club to look after the rats and members in default and so on had taken a day off. The little fellow was unearthed in the photo-etching department, where he examined some half-tone blocks with the eye of a connoisseur. His friend and purveyor was sent for, and then ensued a comical scene. The friend used every argument he could to induce "Joey" to come along home with him, but "Joey" couldn't see it for nuts. At length the court was cleared of spectators, and by and the little chap and his biped friend meandered out of the photo department and into the editorial room. Here he winked (that is, the mongoose did) at our telephone lady, sniffed at the sub-editor, and, dodging the poet, made for the editor of the "Express." Then, mis-liking the glare in the eye of that scribe, he fell to a critical examination of this heating apparatus, and finally wound his way out and into the long passage which, as all men know, leadeth unto the club. It appeared that "Joey" was not at all pleased at the abrupt termination of his holiday and even his best friend kept a respectful distance when coaxing him (home. He had, it appears, been all over the office, printing presses, linotypes, and all, and the wonder was that lie escaped without being turned out by mistake as a special edition."

WRENBURY. CURIOUS ZOOLOGICAL VISITOR.—Last week, while Thomas Davies, gamekeeper to Mr. Walter Starkey, was attending to his duties in Wrenbury Mosses, he discovered a strange animal in a trap. The creature snarled at him, whereupon he despatched it. The animal, which is in the hands of the taxidermist, proves to be a mongoose. Surely this is a curious find in a pre- served English cover?

Wrenbury is in Cheshire.

THE LASCAR IN GLASGOW.

The hunt for the rat-killing mongoose is still very keen in Glasgow, because yet there are no signs of truce the war on rats. Since the ultimatum of the Medical Officer war has been continuous, and rat-catchers have known neither sleep nor leisure. The demand for that Indian weasel, the mongoose, was so sudden that there has been time to meet it with an adequate supply. A "Weekly News" special correspondent the other night accompanied a well-known Glasgow dealer on a round of the shipping just arrived at the harbour from foreign ports. The dealer had many mongoose orders, but never the tail of one to give to anyone. He was therefore very eager to meet the coolie sailor men, who generally burden themselves on their voyages with queer pets of all descriptions for the purpose of selling them on their arrival at British ports. It is nine o'clock p.m. before the first vessel —a big Indian liner—is boarded. A thick fog enshrouds the harbour, and deck of the steamer it; silent the grave. Stumbling in the darkness over ropes and obstructions the fore quarters the vessel are reached. A dusky bearded native with a round cap his head all alone the upper deck. "Mongoose?" shouted ill his car. The native can speak fairly good English, and soon explains that there is no mongoose on board. Any monkeys?" is the next question. "No; no monkeys," is the reply. Asked to the whereabouts His brethren of the crew, the points down a hatchway with an almost perpendicular stair. With muttered objection the inconvenience of seafaring ideas of architecture, the hunting party commit themselves to the stair, and to the black,

Above: *The Chester Courant and Advertiser for N.Wales* Sept 11th 1901.

unknown depths of the vessel. A narrow space leaving just room turn is reached at the foot, with evidently passage anywhere, "Push!" shouts the Lascar from above, at-d a door like a panel in the wall—if a ship may permitted to have walls—gives way, and a dim light struggles out. It is the sleeping quarters of the native crew, and the party pass inside. There are from a dozen to twenty native sailor men, three or four seated round blazing stove, the rest asleep or endeavouring to sleep, in their bunks. The opening of the door the signal for a chorus of coughing, for the night sharp with a bitter frost, and the fog has penetrated even to the...depths. "Mongoose? Monkey?" the query, and half-a-dozen dusky faces are lifted from half a-dozen bunks. The natives round the fire shake their heads, but an alert and scanty bearded Lascar from his bunk takes up the duty of interpreter. "Any mongoose to sell?" is asked. No," replies, rests his elbow. "Monkey?" "No." "Snakes?" "No. snake. nothing." "Not any parrots?"

"No." fruitless errand. There is bird, beast, or reptile board, but there are seats by the reeking stove, even for unceremonious strangers, who come to disturb the rest: of gentle natives. The "Weekly News" correspondent tackles the interpreter in his bunk with questions about snakes and the mongoose. The native explains that has often brought over but not this voyage. told of the war rats, and assured for his comfort that half-a-dozen of the Indian rat killers brought over that particular voyage would have been for him a small fortune. He is not impressed. A Lascar is not easily shaken out of his equanimity,' even when not half asleep. promises, however, to bring some next voyage. " Get mongoose, monkey, Calcutta," exclaims; " bring Glasgow," and it a bargain. Conversation drifts from natural history to general topics. The Lascar admits it cold. He does not like cold. Yes, had seen the Exhibition early in the season, and the Indian Theatre- had much better snake-charming and juggling in India, however. He liked the exhibition. One of the natives at the stove holds two inches of a lighted candle in his hand, while he pores over a little volume covered with Hindustani characters and rude line pictures of animals. The interpreter explains that it a story book, and readily obliges by reading half a page, leaving his warm bunk and coming to the stove to do so. He next translates what he has read into his best English. Strangely enough, the part of the story chosen about a good cat and bad rat, and the pictures the«e animals bear out that there is no deception. The story would seem to bear some resemblance to the English animal stories for children, or Eosope`s fables. A request for some native music is very properly declined by the interpreter, who explains that coolie men are sleeping, that they have to on deck again at five o'clock, and that the tom tom would wake them up. There is, besides, a sick man, very bad with the cold. The tom-tom neither a musical nor a merciful instrument, and it that the coolie is mindful of his sick and sleeping brethren. It a strange sight that presents itself in the dimly-lit and stuffy sleeping quarters. The barefooted, lightly-clad natives round the stove sit with their feet on a heap of coal. From the bunks all round come sounds audible sleep, and there is practically no difference between the snore the coolie and the sturdy British snore. Coughing, too. in Hindustani much same as in English. The scene recalls Kipling's enumeration of coloured seamen resting from the sea"— " Pamba. the Malay, And Carboy Gin, the Guinea cook, And Luz from Vigo Bay.

With a final salaam, and a vigorous and courteous response from the natives, the strangers retire, for other vessels have to be visited. There is no luck this night, however. sailor man has been lucky enough to bring mongoose at the psychological moment, and, although there may be a glut of the animals by the time the vessels return from their next voyage, it is to hoped by that time the extra demand will have ceased. At the next, liner visited, "No mongoose, no monkey" is still the cry, and, as the hour getting late, there is no staying talk with coolies. Another vessel visited resulted the purchase or parrot, and the bird forms the entire bag for the evening. The lascars who make profitable trade in harmless live stock from their own clime, find a ready market London, Liverpool, and other ports. As rule, the officers their ships not encourage a shipfull of monkeys, parrots, snakes, &c., but the Lascars find ready means of bearing their animals safe to port without having them entered in the ship's books. It is no uncommon thing to sec a mongoose running about the men's quarters, a couple of monkeys screeching out tho captain's hearing in the hidden recesses of the fo'c'sle. The Lascars kill sheep for their own consumption on the voyage, in accordance with the dictates of their caste, I and monkey and mongoose, especially the latter, feed fat the mutton. But this a caste secret. In port,

when the dealers fail them, or the lascars are too eager for a bargain, they bear their animals to private houses, and in Glasgow it is no uncommon thing bee a Lascar with a mongoose squatted on his shoulder, or a young monkey curled under his jacket, on the search for purchaser. Nor does the trade in live animals limit the commercial enterprise of the Lascar. He brings also corals and trinkets, rugs, and native curios, and evidently finds a market for these things, although that market has been almost destroyed in Glasgow since the plague scare of more than a year ago. and the Lascar has scarcely the heart brine anything the port now.

Another suspected Indian mongoose was sighted in Northamptonshire in September 1904. According to the *Northampton Daily Record* (?) September 28th 1904:

"A Rare Animal. Whilst out rabbit-shooting on Tuesday last week, at Mr Banks` lime kilns, near Moulton Park, Mr W. Parbery, of the Old Five Bells Kingsthorpe, was fortunate enough to secure a rare natural history specimen. There was a good deal of speculation as to what the animal was but through the kindness of Mr T.J. George and others it has been found out to be an Indian mongoose. The animal, when "bolted" by a ferret (which was muzzled) showed fight, but was soon despatched. It is supposed that the specimen (which is being preserved) had escaped from confinement."

Jumping ahead to 1906:

"WANTED, A MONGOOSE
A Slough correspondent telegraphs that the master of Eton Workhouse yesterday laid before the guardians a huge pile of letters

Animals & Men — Issue 54

which lie had received from all parts of the United Kingdom from people anxious to assist him in ridding the workhouse of rats. Amongst the letters was one from the wife of one of the best-known English bishops, in which she stated that amongoose had quite cleared the bishop's palace of a similar plague. Another remarkable suggestion was the employment of a guinea-pig, which the writer of one of the letters said he had found most successful. Patent poison and trap vendors and professional rat catchers galore offered their services, and the guardians left it to the master to make a choice." *Evening Express* Apr 4th 1906.

"FARMER TRAPS MONGOOSE. A farmer living at Rond, near Trowbridge, set traps recently to catch rats near a rick in has field. Next day he was surprised to find a strange animal, about the size of a cat, in one of the traps. It was of a very savage nature, but the farmer took the animal home, and a gamekeeper was called in to put it out of its misery, as one of its legs was nearly amputated. The animal was sent to a naturalist, who said it was a mongoose." Evening Express 17 Sept 1907

Then in 1926 came a story from Gloucester:

"RIKKI-TIKK"
THE MONGOOSE AS A DOMESTIC SCOURGE
RECENT CAPTURE AT HEADINGLEY IS THE MYSTERY OF A STRANGE VISITOR DISSIPATED?

By "An Old Poacher".

I don't think we will hear anything more of the "weasel" or "stoat" or other blood-sucking animal that caused such consternation in home in Leeds month so ago. A week last Friday was on my way to fill a professional appointment as rat catcher in the close vicinity the railway, not five minute's walk from the house where the trouble has occurred, when my dog started some small animal, which bolted into a hole. Not got getting a proper look at it I imagined it might be a stoat or weasel. I pulled out of my pocket a well-trained ferret but, to my surprise, immediately I put in the hole it turned round and began 'to "sis." or what one would call in a cat to spit.

MONGOOSE CAUGHT.

Believed to have been responsible for mysterious attacks on poultry roos's, an Indian mongoose, 18 inches long, has been caught at Hedon, East Yorkshire.

The *Yorkshire Evening Post* 1 May 1912 p. 3

I put a fuse of gunpowder in the hole, and once out bobbed the creature. It was young mongoose. I trapped it alive. Since that date, understand, they have had no further trouble the house at Headingley. The mongoose was a fine specimen, but very young, and had evidently escaped from somewhere. I kept it until yesterday when I sold it for 25s...

The *Gloucester Citizen* 17 March 1926 p. 8

And ten years after that:

Derby Daily Telegraph
19 February 1936 p. 10
MONGOOSE CAUGHT FOUND IN BUNCH OF BANANAS AT DERBY

A live mongoose has been captured on the premises of the Empire Banana Company, Derby. The mongoose, which is about six inches long, was discovered in a bunch of bananas from East Africa. Little difficulty was experienced in catching it. Mr. E. Nield. the general manager, told a "Telegraph" representative today that he was keeping the animal in cage, with a view to taming and training it. At present it is fed on milk. The mongoose is used extensively abroad for destroying insects and reptiles, on which it feeds.

In a blog dated June 27th 2015 Malcolm Peaker wrote:

"Mongooses are used in London warehouses to keep down the rats. One big pet shop reports that they sell at least six of these animals every month for this purpose." [3] This was in the December 1938 issue of *Animal and Zoo* magazine.

It is interesting to note that the years when most of these reports were received, as stated in the opening paragraphs to this paper, were the ones when human interaction between Britain and British India was at its height due to the improvements in both transports and communications, which meant that it was far easier for the Mothercountry to keep in touch with Her Colonial Possessions.

It is also interesting to note in the 1926 story from Gloucestershire, reference to Kipling's seminal story *Rikki Tikki Tavy*, which was first published in book form in 1894 and features the adventures of a young and heroic mongoose. Possibly as a result of the popularity of Kipling's Indian stories mongooses were sought after pets in the UK at the time, and - for example - can be read about in Gerald Summers' *The Lure of the Falcon*, which was written in the mid 1970s and described the author's boyhood as an amateur naturalist in the 1930s.

In the early 1990s the popularity of a Hollywood movie about crimefighting turtles led to the popularity of these animals as pets, and the subsequent release of many unwanted turtles into the British ecosystem. The events described in this paper suggest that this was possibly not the first time popular culture influenced the British cryptofauna.

References.

1. `Proposed importation of mongoose to exterminate rats`. April 16th- June 24th 1929 CO 78/184/6
2. Malcolm Peaker. The Isle of Bute Mongoose of December 1900. Zoology Jottings June 17th 2015. http://zoologyweblog.blogspot.co.uk/2015/06/the-isle-of-bute-mongoose-of-december.html
Malcolm Peaker. More on Mongooses Employed in Britain As Rodent Catchers. Zoology Jottings June 27th 2015 http://zoologyweblog.blogspot.co.uk/2015/06/more-on-mongooses-employed-in-britain.html

DISCUSSION DOCUMENT: DUTIES FOR REGIONAL REPRESENTATIVES

The CFZ had had regional representatives for over twenty years now. Some of them have done remarkable things, some nothing at all, and some something in between. I originally intended my first wife to manage the list of regional reps, but as history shows, that never happened.

Ever since Alison and I split up I have been intending to ask someone else to take over the job, and finally a few months ago I got around to it. Ronan Coghlan has agreed to take over the onerous task, and has come up with a list of suggested roles

for regional representatives, which I post here for public discussion.

[1] In the event of a reported sighting of a mystery animal in the representative's area, all possible data should be gathered and forwarded to CFZ. Likewise, news of further developments should be sent on as they occur.

[2] Representatives should try to discover if there were any sightings or other anomalous events in their areas in the past, but should only send on stories of UFOs or ghosts if they consider them important, as otherwise their is the danger of CFZ being swamped.

[3] Representatives should, if possible, look into local folklore to discover if stories of anomalous events in the area occur. Liaisons should be initiated with the Bird, Butterfly and Conservation Officer in their areas where possible. They should, in addition, try to gather an archive of Fortean zoological material from their local studies libraries.

[4] Representatives should initiate liaisons with groups dealing with anomalies and nature in the area, provided they consider them and their personnel suitable.

[5] Representatives should have the option of offering sales of books to local bookshops. However, some might find this distasteful and so this should not be regarded as an actual representative's duty.

The editor and his compadres welcome letters for publication on all subjects covered by this magazine. However, we would like to stress that neither this magazine, or the CFZ are responsible for opinions expressed, which are purely those of the letter writer.

Walk on Gilded Splinters

Dear readers,

At the end of May 2015 I found the following story in the Hong Kong Police Magazine for March 1956, volume 6 no. 1 whilst at the British Library in London. This was a magazine which came out about 6 times a year up until at least 1970 and which reported in English and Chinese on events pertaining to the police, such as crimes, historical matters, unusual events and reports from different regions around Hong Kong, such as Eastern Hong Kong, which is where this event of cryptozoological significance took place. It was most likely another Hong Kong giant salamander. There is only one definite giant salamander case from Hong Kong, which I wrote about in issue # 53 of Animals and Men, that is the case of *Megalobatrachus sligoi* in 1922.

I quote directly from the Hong Kong Police Magazine of March 1956 page 39. The capitals are in the original version.

"At the time of writing this Divisional letter, the main topic of conversation is of a recent heading in our Report Book which reads "Unknown Amphibious Creature found in Eastern Police Station Compound" .It appears that at about three o` clock one morning our staunch new arrival, whose name I shall not remind you of, decided he needed a breath of fresh air, to relieve his troubled mind of the second night I.O.D`s (Inspector on Duty?) duties. In order to do this, he proceeded to the rear of the Charge Room and thence to the compound. As he was filling his lungs with the fair fragrance of Wanchai, to his horror he beheld the THING. At first he could not believe his eyes, but after a thorough rubbing and quick face slap, he perceived that the THING was still there, and in fact IT was moving toward him. He was rooted to the ground as he watched IT slowly advance upon him.

It had emerged from a drain about ten yards away, was bout eighteen inches long,had four legs, a tail and a spade head with a mouth full of saw-edged teeth, or at least so the varied stories circulated later said.

Shaken to the core, but still mindful of his stature and position, he summoned support from within the Charge Room. The support proved varied in their opinion as to whether the creature was a dragon or a baby crocodile and in the end our hero had to manoeuvre IT into a large oil drum on his own.

The following morning the THING, now an

Inscribed on the front in pencil: 'M. maximus' and 'M. sligoi', the three figures of giant salamanders are shown in this drawing, two in side view and one seen from the ventral surface, the drawing has not been signed or dated so it is a bit of a mystery. It may have been prepared to accompany a paper entitled 'On a new giant salamander, living in the Society's Gardens', which was read by E.G. Boulenger to a meeting of the Zoological Society of London (ZSL), when the paper was published in 'Proceedings of the Zoological Society of London', 1924, pp. 173-174, it was not illustrated and no reference to this painting was made in the paper.

object of curiosity, was closely examined. It was about eighteen inches long and can best be described as a large tadpole in its later stage of development.

It was subsequently removed to the Hong Kong University, where it was later identified as a `Giant Salamander`. Many of these are evidently imported into the Colony for sale in the local markets. As a matter of interest, they are perfectly harmless and do not possess saw-edged teeth.

However, I feel sure that those who have done night duty, and who among us have not, can fully appreciate the horror of being confronted with one of these creatures at such an hour in the morning."

Best Wishes
Richard Muirhead

ANOTHER SLIGO'S SALAMANDER

Proving, as if any proof were necessary, that the world really is a peculiar place look at this. **Sligo Creek Elementary School** 500 Schuyler Road. Silver Spring, MD 20910 has an animal on its logo. And can you guess what this animal might be?

Hardcover: 480 pages
Publisher: Bloomsbury Natural History (4 Jun. 2015)
ISBN-10: 1472924509
ISBN-13: 978-1472924506

I really don't know how many books I have read over the past few years that mention the 1970 Isle of Wight Festival. It was the show at which Mick Farren and a couple of mates, under the guise of the UK White Panther Party put together an anarchic free festie on a hill overlooking the official event.

It was the show at which ELP are rumoured to have got together for a jam with Jimi Hendrix, prompting nearly half a century of rumours that have categorically been denied by Greg Lake at least. It was one of the last shows Hendrix ever played before his death eighteen days later. And it was the show that a young Matthew Oates attended because he wanted to see Leonard Cohen.

For those of you who have never heard of him, Oates is somewhat of a legend amongst those of us who are interested in the long twisted saga which surrounds the sixty odd species of butterflies which are found in the United Kingdom. He is the National Specialist on Nature for the National Trust, and his biography on their website proclaims: "Butterfly expert, author, poet…Matthew Oates is something of a Renaissance Man. Celebrating 50 years of butterflying in 2013, Matthew is one of those rare ecologists with a background in the arts – his passion for butterflies matched only by that for the great English poets Coleridge and Edward Thomas.

Graduating in English, Matthew then moved into the world of nature conservation and has been at the Trust since 1990. He is particularly drawn to people's relationships with nature, places and seasons, and increasingly the impact of weather on wildlife.

Matthew is well known to the media. He's made a number of appearances on BBC Radio 4 - from the *Today* programme and *Shared Earth*, to presenting two short series: *In Pursuit of the Ridiculous* and *In Pursuit of Spring*. His TV credits include *The One Show*, *Springwatch*, *Great British Summer* and *Butterflies - A Very British Obsession*."

This book, very thinly disguised as an autobiography is nothing less than a personal look at fifty years of British butterfly reports from 1963 to 2013, based around, but not inclusive of, his own observations across the years.

He is what Bob Marley once called a "natural mystic" and his prose and poetry reflect a deep, spiritual and completely overwhelming love of the British countryside and its papilonid inhabitants, of the sort that one found in the reminiscences of 19th century country parsons, but is increasingly uncommon in our own degenerate age.

I have always been fond of natural history memoirs, the moth collecting books of P.B.M Allan - a trilogy: *Moths and Memories*, *A Moth Hunter's Gossip* and *Talking of Moths*, being particular favourites - but until a few years ago I thought that this was a literary genre which had vanished forever. Then along came a book by Patrick Barkham, which resurrected the genre, but also managing to bring it up to date with such 21st century additions as text messages, soon to be ex-girlfriends, and all sorts of other things that dear old Philip Allan would probably never have mentioned (although one of the trilogy listed above does hint at a teenage dalliance with the massively saucy daughter of a country innkeeper).

Now Matthew Oates comes along with probably the most poetic and romantic (in the literary sense) book on British butterflies, complete with comments about Bob Dylan's *Blood on the Tracks*, which unlike so much poetry that I have read by scientists, is truly not at all bollocks! His prose even borders on the Richard Jeffreysesque, and - trust me - that is truly high praise indeed as far as I am concerned.

The tragedy of this book is that it is unlikely to be read outside the butterflying community, and that is a great pity. This is the sort of book that should be read by anybody who appreciates the countryside and the world about us, and also contains a fair smattering of social history of the British middle classes, chronicling a way of life that is unlikely ever to happen again.

I truly recommend this book to everyone who reads this magazine, even the sort of person who would never consider buying a book about little fluttering insects.

Well done Matthew. JD

Paperback: 142 pages
Publisher: CreateSpace Independent Publishing Platform; 1 edition (20 April 2013)
Language: English
ISBN-10: 148417173X
ISBN-13: 978-1484171738

What do you get if you take M.R. James, add a pinch of *Tales of the Unexpected* and *Blood on Satan's Claw,* sprinkle in some Nigel Kneale and stir it up with some choice E.F Benson?

The answer is Chris Lambert's unique book *Tales from the Black Meadow*. Chris is a drama teacher whom I first met when he was presenting a scene from his stage version of *Night of the Living Dead* performed by his pupils. Would that I had teachers like that. Who needs Shakespeare when you can have George A. Romero?

Tales of the Black Meadow purports to be a collection of folk tales, legends, rhymes and stories from a remote and foreboding area of the North Yorkshire Moors. Folklorist Rodger Mullins of the University of York went missing in 1972 leaving no trace but a collection of work pertaining to The Black Meadow.

The Black Meadow seems to be a shunned place bypassed by time. It reminds one of H.P.Lovecraft's Dunwich or one of the desolate areas of the East coast in which M.R. James set his disturbing stories.

The stories and rhymes contained in the book are notable for both outstanding oddness and the queer ring of 'truth' that marks out many a good story. The anecdotes remind one of some of the weirder stories covered within the pages of *Fortean Times.*

One of the best stories, *The Land Spheres*, involves black spheres that emerge from the mists one night and simply devour all light. The story put me in mind of an account from WWII when a group of night watchmen saw a black sphere emerge from the night and seemingly toss about some heavy railway sleepers. The story isn't strictly horror but it is so deeply strange as to be unsettling. It is this kind of story, both on the printed page and in real life that I find by turns the most intriguing and the most disturbing.

There are many more stories of 'high strangeness'. *The Watcher from the*

Village sees a small community thrown into paranoia by a black *thing* that simply watches. In *The Fog House* a lost traveller stops at a house of fog and is guested by a man of fog. He eats food of fog and drinks wine of fog whilst sat in a fog chair at a table. He lays his head in a fog bed and sleeps with a girl of fog.

The Long Walk to Scarry Wood is an effective little tale simply told and all the more horrid for it. It is just the story of a man on an errand on a road that goes on forever. Time and space are bent in the Black Meadow. This is pastoral horror from the same mould as *The Wicker Man* and *Night of the Demon*.

The book itself is accompanied by two CDs. The first consists of eerie music to attend each tale. It is the second CD that is of most interest though.

This is a pseudo-documentary about the Black Meadow and the disappearance of Professor Mullins and other researchers. It has interviews with his colleagues and others who have had cause to spend time in the area. In one part we learn of a Yorkshire TV children's programme *Children of the Black Moor* (no doubt inspired by the series *Children of the Stones*) filmed on location. One of the child actors never returns.

It's done in the same vein as *Ghostwatch* and *Alternative 3*. I have no doubt that some readers will come away from *Tales from the Black Meadow* thinking it is all real.

How long I wonder until some of the motifs from the book find their way into actual folklore via repetition and social osmosis? RF

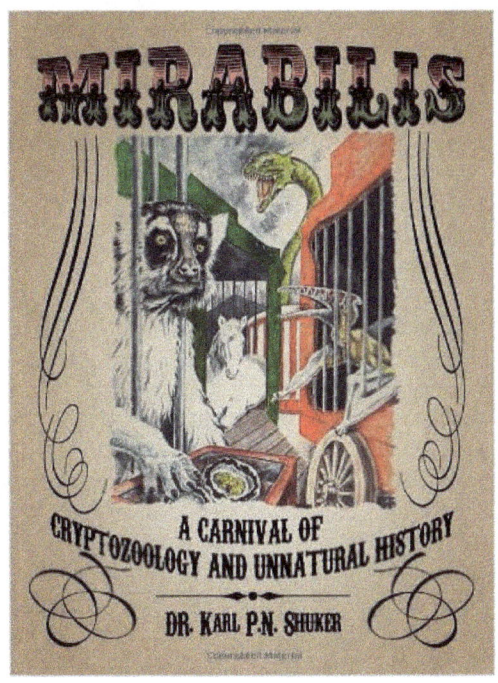

Paperback: 194 pages
Publisher: Anomalist Books (12 Aug. 2013)
Language: English
ISBN-10: 193839805X
ISBN-13: 978-1938398056

One of the most prolific and entertaining fortean writers of modern days, Dr Shuker never fails to delight and impress. Mirabilis is no exception. Its cover, wonderfully rendered by Ant Wallis, brings to mind the classic film *The 7 Faces of Dr Lao* based on the novel by Charles J Finney.

In old circus-like cages a giant lemur, a unicorn and a pterosaur are seen, whilst behind them a dragon looms.

Karl has a wild talent for covering

subjects most other authors in the field ignore, have forgotten or have never even heard of. He digs up new details and cases like a pig digs truffles. Case in point is the chapter on 'Trunko' the weird, white furred, elephantine trunked beast said to have been seen battling two whales off the cost off Margate, South Africa in 1924, subsequently washing ashore dead.

The beast had baffled cryptozoologists for decades. Together with his German colleague Markus Hemmler, they piece together the long overdue answer to the riddle with the help of never before published photographs uncovered by Hemmler.

Elsewhere Karl looks at the fantastic fauna of Madagascar turning up strange stories that escaped even Dr Bernard Heuvelmans, the grandfather of cryptozoology himself. Here are sightings, some intriguingly recent, of what sound like giant lemurs thought long extinct, and pigmy hippos, but also what sound like primitive hominans lurking in the wilder parts of this island continent.

As a lifelong reptile keeper and self-confessed crocodile fanatic the chapter on crocodilian mysteries was bound to be a favourite of mine. The stories of giant crocodiles in Africa and huge marine crocodiles were familiar to me.

Less so the horned crocodiles of Madagascar and the 'frog' crocodiles of Borneo. Other oddities include the strange case of the New Guinea penguins, antlered sea snails, giant spiders in suburban Britain and monster tortoises. Like a literary cabinet of curiosities this book has something to tantalize every reader. *Marabilis* - wonder indeed. RF

Paperback: 174 pages
Publisher: Skylight Press
(28 Feb. 2015)
Language: English
ISBN-10: 1908011904
ISBN-13: 978-1908011909

During the summer of 1996 a friend of mine and her boyfriend were having a picnic lunch near Postbridge on Dartmoor. She was momentarily distracted from her cheese and tomato sandwiches by the sight of something small and brown scuttling around on a small pile of stones half obscured by bracken. She was a keen amateur naturalist and immediately thought that it might be a weasel, so – quietly – she got up, and went to investigate. She was not at all prepared for what she saw.

It was – she told me, a few days later – a little man about 6 inches high. He appeared to be completely naked except for a large, wide brimmed hat, and was a deep chestnut brown colour. He was busily engaged in moving a small pile of pea gravel, which – of course – to him was sizeable boulder. She let out an involuntary gasp, and the Homunculus gave a start and looked straight up at her. They stared at each other for several seconds before the little man dropped his burden, hissed audibly and scuttled off into the undergrowth.

She had encountered a fairy. No, I am not making this up, and – I'm convinced – that neither was she. Yes its true, even in our increasingly urbanised and technological environment, people do encounter beings from fairy land.

I think that it is quite possible that many people on first picking up this book, will naturally assume that it is a book about the well known, and culturally ubiquitous fairies of childhood lore; little people with gossamer wings who flit through the undergrowth doing their own inimitable thing.

This would be a perfectly reasonable supposition, but it would be completely wrong. There are, of course, "conventional fairies", if indeed one can use the word "conventional" to describe something which is firmly from the realm of the supernatural. But it is so much more.

This is an exploration of what Ted Holliday described as "The Goblin Universe", only Coghlan goes a long way towards explaining a reasonably cogent paradigm by which these things exist. I used the work "supernatural" earlier in this review, but it is a term I hate.

Like "paranormal" it is a complete misnomer; these things are perfectly normal, and perfectly natural. It is merely that our society, and more specifically our scientific community doesn't know how to categorise them, and doesn't – as yet – understand how they work.

I am glad to see that Ronan Coghlan agrees with me.

I am glad to see that this book also covers a wide range of other entities from the realm of faerie. Things such as: puca, owlbeings, gnomes and other earth people, and even kelpies and water horses.

This is an excellent little book because not only does it give thumbnail sketches of a whole range of different encounters with other worldly beings, but in doing so gives a remarkably concise overview of a whole range of different phenomena which do seem to be somehow related to each other.

This is a remarkable book and Coghlan should be congratulated on the way that he has successfully combined scholarship with story telling.

The only draw back is, that knowing him well, there are times in the book that I hear him delivering one of his droll asides in his own inimitable manner, and laugh aloud. However, as I am a bear of rapidly diminishing brain, I then lose my place and have to start all over again. But it is always worth it. JD

This year was the sixteenth Weird Weekend, and it truly seems unbelievable that an event that a few of us thought up for a laugh back in 1999 is not only still going, but is going from strength to strength.

Over the years it has mutated and changed several times, and the most recent of these sea changes took place last year when we were unceremoniously booted out of the community centre in my home village where we had held the event for eight years. I don't want to go through the details of this again, because they were both upsetting and unpleasant and if you really want to know, a cursory search online will enlighten you. Anyway, it came to pass that last year we changed venue at the last minute to The Small School in Hartland; a radically alternative seat of learning established by Satish Kumar 30 odd years ago. It is an organisation far more in keeping with my personal code of ethics than the increasingly commercialised and capitalist ethos of our previous venue.

The arrangements for this year's event went surprisingly smoothly, in fact they went *too* smoothly. The only thing that actually went wrong was a whole string of people upon whom I had been banking were unable to attend due to a succession of unforeseen family commitments. In fact, statistically it felt like there were more than usual, but I have a suspicion that this is my innate paranoia speaking and that actually the drop out rate was much the same as usual. No doubt this will happen again in the future and I shall once again get all paranoid on you all.

As always, it all kicked off on the Thursday evening with the now legendary CFZ cocktail party; luckily the events of 2008, when we had 120 people dancing riotously to a disco on my lawn, were not repeated. On that occasion we had only planned for a fraction of that number and had been overtaken by events. The fact that the CFZ crew were mostly in an inebriated state didn't help, and I vowed never to allow such a Bacchanalia take place on my premises again. This year's event was much more civilised. We even had live music from Dogleg, Stargrace, and our very own Davey Curtis. I've been wanting to have live music at the Weird Weekend cocktail party for years but this is the first time we ever got around to it, and I think it was a great success. Certainly, it is something we shall be doing again.

YOU NEVER HAD IT SO WEIRD

The event proper kicked off on the Friday evening with Nick Wadham's Wild and Deadly Animal Show, which - as always - delighted the youngsters in the audience, and gave us oldies something to think about.

He was followed by Lee Walker who is – against some stiff opposition – probably the best and most stylish writer that we have on CFZ press. He was launching his new book; *Glimpses in the Twilight* which is his second collection of astonishingly macabre vignettes of a boyhood spent investigating the nastier urban legends of Merseyside. He didn't disappoint and – as always – was a great success with Weird Weekend attendees.

Finally there was Lars Thomas doing a talk that I have wanted him to do for at least three years, but – because of the slings and arrows of outrageous fortune – had always fallen by the wayside. It was a talk on Microcryptozoology; a neologism invented by Lars to describe the cryptozoology of small creatures, often invertebrates – that are all to often ignored by devotees of cryptozoology in general. It was a fascinating talk which included a creature only known from one specimen, a corpse eating fly that disappeared for hundreds of years, and various insects – quite probably new to science – which are described elsewhere in this current volume and which Lars captured whilst on the recent Tasmania expedition.

The evening ended with the now traditional bedtime story from Silas Hawkins who read from one of Richard Freeman's collection of horror stories.

Day two dawned bright and shiny and kicked off with an open ended discussion chaired by me and Richard Freeman about the nature of cryptozoology. We then introduced everybody to Carl Marshall who was kind enough to

donate a very special specimen to the CFZ collection, It is a marten probably taken in Dorset during the 19th century. Twenty years ago I published a book suggesting the existence of a number of cryptozoological animals in the western counties of the United Kingdom. Two of these are of particular interest here. I suggested that the pine marten (*Martes martes*) had not been hunted to extinction in the late 19th century as was then believed. Time has proved me broadly right. My second claim was that another species, the beech marten (*Martes foina*) which is found alongside the pine marten over most of its range, is actually found in the UK.

We believe that this stuffed marten may be of the latter species and therefore an irreplaceable cryptozoological specimen and Carl, Lars and I explained the progress of our investigation into it.

Next up was Steve Rider who gave a fascinating talk about the difficulties of obtaining photographic evidence of 'paranormal activity' in a digital age.

The next talk was by far the most controversial of the weekend. Jaki Windmill and I have been friends ever since I met her singing with the late Mick Farren I booked her to talk on Astroshamanics because I'm very fond of her and I find the subject mildly interesting, although I will be the first to admit that I didn't believe a word of it. However, in the spirit of scientific investigation I joined in her workshop, followed her instructions, and found myself in the most altered state that I have ever been without chemical intervention. This is definitely something that I will investigate further into the future. She was followed by Richard Freeman talking about dragons, a subject on which he is undoubtedly one of the world's most leading experts. Richard was followed by my old friend Judge Smith, the original drummer with *Van Der Graaf Generator* talking about the history of Ouija boards. It is a rich and fascinating subject about which, until then, I knew very little. Well done Judge.

We then had music from Jaki, the CFZ awards which, this year, were won by Helen Taylor, Danny Owens, Lee Walker, Shoshannah McCarthy, Jacquie Tonks, and then legendary Adam Davies spoke about his various expeditions around the world in search of manbeasts. Finally, Saturday ended with another appearance by Lars Thomas giving a report of this year's expedition to Tasmania in search of the Thylacine.

Sunday started off with a presentation from my old friend Richard Muirhead; we were friends as children in Hong Kong and I have known him since he was six years old. For the last two decades we have been intermittently working on a book on the mystery animals of our childhood home. Richard gave a fascinating presentation detailing some of the stories which he has uncovered over the years.

He was followed by Rosie Curtis, a young lady whom I have known since she was a toddler and who is a very dear member of my adopted family. She was talking about the peculiarly 21st century phenomenon of Creepypasta; gruesome or scary stories that have arisen concerning computer games which have then become internet memes.

Our next speaker was somebody with whom I have been corresponding for about twenty years, but whom until this weekend I had never actually met. Rob Cornes has been working on an exhaustive book about the long necked seal hypothesis to explain sea serpent sightings for many years now, and in his talk he presented what was probably the most well-argued and cogent hypothesis that I have yet seen.

The next talk was from Shoshannah McCarthy who, as well as being my stepdaughter is a highly qualified vet. She gave a fascinating presentation on the subject of feral cats both from a Fortean and a biological perspective. It was a fascinating talk, and very well received.

Then came the irrepressible Ronan Coghlan who gave an enthralling account of his adventures searching for the semi-mythical Irish master otter and other Hibernian cryptozoological conundra. Finally, an overweight hippy with a bad hip and a worse attitude took the stage, and it was all over for another year.

I would like to say here, by the way, a big thank you to my assistant Jessica, and my wife Corinna but for whom I really could not have done any of this.

Once upon a time I had a team of over twenty people, but due to natural wastage and a whole slew of other circumstances, we now have only a fraction of that, and Corinna, Jessica and I bore the brunt of the preparations. I would also like to thank my nephew Dave Braund-Phillips and his brother, my Godson Greg who did the AV. Dave has been doing it ever since 2006 and sadly this may be his last year but, plans are afoot for next year, probably involving Jessica, Danny and Greg. I would also like to thank Nick Wadham for the loan of his AV equipment, the Phillipson family for having Ronan as a lodger for the weekend and the Rider family for doing likewise with Lars Thomas and his son.

The biggest thank you of all, however, has to go to the ladies and gents of The Small School who did a remarkable job with food and drink for the whole weekend. See you next year.

We publish a lot of books. Indeed, I think that we could quite easily claim to be the world's foremost publishers of books about Fortean Zoology and allied disciplines, and our Fortean Words imprint is doing a great job in producing books on other non-zoological esoterica. However, I feel that it would be unethical to review our own titles. So here, to end this edition of *Animals & Men*, is a brief look at the books we have put out so far this year.

Paperback: 324 pages
Publisher: Fortean Words (28 July 2015)
Language: English
ISBN-10: 1909488348
ISBN-13: 978-1909488342

"Let me take you down, coz were going…" to a place that you may think you know quite well but the city of Liverpool is a far stranger place then any of you could imagine.

Lee Walker is a paranormal researcher, auteur, and one of the best writers on the subject operating in the world today. This is his second collection of urban legends and high strangeness from the haunted heart of Merseyside. Elegantly macabre and often down right terrifying the stories in this book raise the bar for anybody working in, and writing about, esoteric subjects and cannot be recommended highly enough. Walker's Merseyside is a strange and unsettling place complete with giant crabs, brutal and disturbing fairy stories, and even a surviving Nazi war criminal. Strange days indeed; most peculiar Mama!

Paperback: 370 pages
Publisher: Fortean Fiction (16 July 2015)
Language: English
ISBN-10: 190948833X
ISBN-13: 978-1909488335

Stray into the woods and forests and you will enter into another world; a world of creatures that live by their own rules, protect their own kind with fierceness, and view all strangers who venture under the protective boughs with deep suspicion. Tread the forest floor with care, for this is also the home of the hairy man - the wildman of the woods. Is he a man? Is he a beast? Is he something in between? Whatever he is, he is spoken about by humans in hushed voices. They are scared of him and they tell tales of him to their children to scare them from entering the hushed darkness of the tree kingdoms. They call him many names; the woodwose, the wudawasa, the wodwos amongst them. The bane of a high-born daughter takes her unintentionally through such a forest on her last journey as a single woman to wed the man of her father's choice. Imposters from another land tramp through such a forest on a mission of their own, killing everything that comes across their path. The lives of some of the creatures that dwell in this place become unavoidably entwined with both these trespassers. The lives of some will change. Some will cease completely.

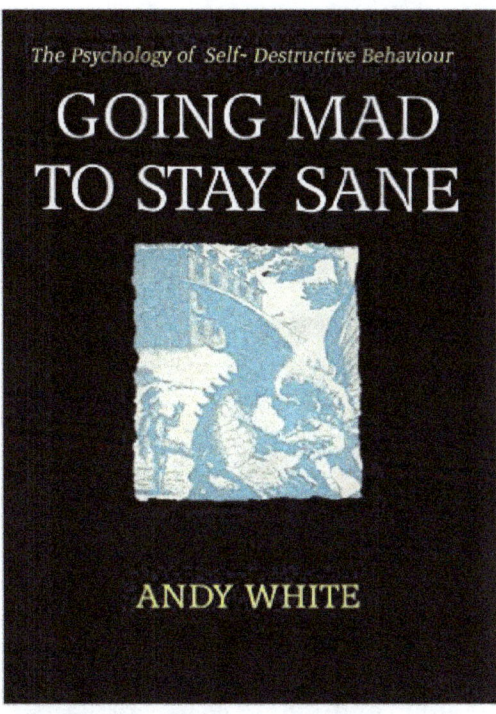

Paperback: 96 pages
Publisher: CFZ Communications (14 July 2015)
Language: English
ISBN-10: 1909488321
ISBN-13: 978-1909488328

Self-destructive behaviour has traditionally been viewed in an entirely negative light. As a result, attempts are constantly made to 'fix' it without asking what its actual purpose may be. Going Mad to Stay Sane invites us to rethink our attitudes. It sets out to understand the soul's purpose in visiting violence upon itself; substance abuse, compulsive sexuality, obsessive dieting or the grandiose hauteur of a superiority complex all come under scrutiny. In analysing its roots and its manifestations, the author asks us to consider the possibility that the impulse to visit violence upon oneself may be the only means available for the soul under siege to preserve itself and state its distress. Self-destructiveness is a notoriously difficult phenomenon to bring to healing, not least because the various schools of psychology have such partisan attitudes towards it, approaching it from within the narrow parameters of their chosen theories. This book, rather than arguing for one perspective or another, finds a place for them all within the compass of a mythical tale: the story of King Midas, who wished for everything he touched to be turned to gold. Through the tale of King Midas, Andy White shows how our self-destructive urges can also point the way to our salvation. Andy White was born and brought up in Africa. He trained in London, practised as a psychotherapist for many years and now lives in North Devon as a writer and artist. www.andywhiteartist.com

In 1958, in the post-Stalinist political thaw, the Soviet Academy of Sciences diverted itself for a time with the exotic and sensational subject of the Himalayan yeti. As the Academy had received reports of similar creatures in the mountains of Soviet Central Asia, it set up a special commission to collect evidence on the subject and launched a major expedition to the Pamirs to establish the existence of snowmen there. The expedition was a failure (this book explains why), and this put an end to official interest in the matter. Snowman studies (or 'hominology,' to give its modern term) was declared by the academic establishment to be a pseudo-science, along with astrology and parapsychology.

Paperback: 260 pages
Publisher: CFZ
3rd edition (29 Jun. 2015)
Language: English
ISBN-10: 1909488305
ISBN-13: 978-1909488304
Russia - or to be exact, the Soviet Union - was the first country to probe the snowman riddle on a scientific basis.

Paperback: 192 pages
Publisher: Fortean Fiction (18 May 2015)
Language: English
ISBN-10: 1909488283
ISBN-13: 978-1909488281
We cannot resist the call of the sea. Enter a strange world where things are never quite what you expect, in this collection of science fiction and weird tales from Kate Kelly. Ancient ruins, lost civilisations, alien visitors and

restless ghosts; sinister technology, environmental disasters, dark secrets of the past and the mysteries that lurk beneath the ocean waves... There are tales here that blend the past with the present, explore the possible futures we are creating and touch on the darker side of the world we think we know. Kate Kelly is a marine scientist by day and a writer by night. Science fiction and weird fiction are her favourite genres, often inspired by her fascination with the sea. She loves writing short stories enjoying the liberation they afford for exploring new ideas and forms. Several of the tales in this collection were previously published to critical acclaim or have been shortlisted in major competitions such as the Yeovil Prize, where Kate was twice winner of the Western Gazette award for best local author. Her first novel, Red Rock, a Cli-Fi adventure for young adults, was published in 2013 by Curious Fox. Kate and her family live in south-west England, near the coast and when she is not writing she takes to the sea in her kayak, or can be found wandering the remoter stretches of the South West Coast Path.

Paperback: 212 pages
Publisher: Fortean Words (18 May 2015)
Language: English
ISBN-10: 1909488291
ISBN-13: 978-1909488298

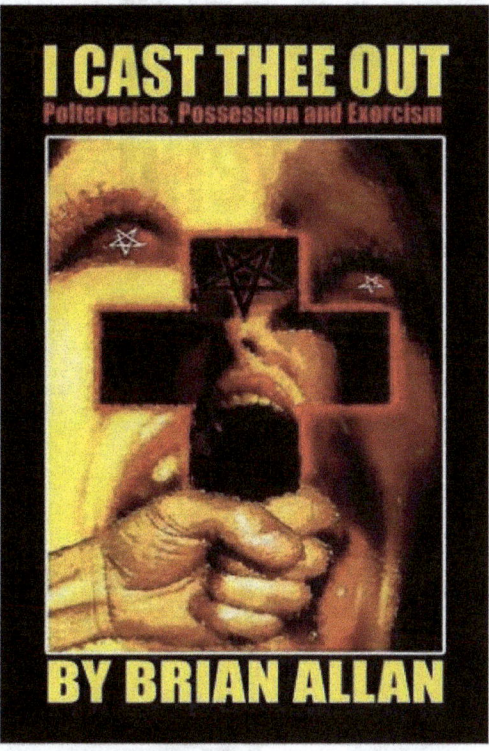

In the course of researching material for this book I was confronted with a multitude of almost unbelievable and differing nuances of opinion and belief when dealing with the subject of demonic or Satanic possession. These took the form of everything from a simple faith in saints and demons espoused by the early, especially medieval Christians, to supposed vampires and the largely self-appointed slayers who take it upon themselves to rid us of these supernatural hybrids. Since vampirism is also arguably a form of possession by an evil spirit, it dovetails easily with the more conventional model. These beliefs acted as a kind of two-way filter, both feeding and evolving into the infinitely greater sophistication and borderline charlatanry of what passes for Christianity in some supposedly Christian schisms in the modern world and there are more of these than one might think. As far as those who become possessed and are deemed to be in need of exorcism go, there is no defining personality type, because the individuals who display the traditional signs of apparent demonic possession can be either devoutly religious or have no religious affiliations whatsoever. As with mental illness the person may seem absolutely fine on the outside, but this gives no clue as to what is going on inside. This has been likened to a moving vehicle where externally all appears as it should be, but the observer has no idea who (or what) is driving it; usually until it's too late, which is why mental illness and possession often appear virtually identical in nature. To be sure, there are still many extremely pious and genuine people out there who successfully challenge perceived evil spirits and by their lights wrestle 'that old serpent' Satan to the ground on an almost daily basis. However, there are also individuals and groups of individuals who, although loudly and vociferously proclaiming their scriptural credentials, have an altogether much darker agenda which, they claim, is also founded on biblical principles.

It is among these groups that we will find the most appalling, shameless, devious and dangerous hypocrisy imaginable.

weird weekend 2016

19-21 August 2016
Three Days of Monsters, Ghosts and UFOs

The Small School, Hartland, North Devon

YOU'VE NEVER HAD IT SO WEIRD

www.weirdweekend.org

www.ingramcontent.com/pod-product-compliance
Lightning Source LLC
Chambersburg PA
CBHW071325040426
42444CB00009B/2083